Gadabouts and Stick-at-Homes

by Lorus and Margery Milne
illustrated by Sarah Landry

Gadabouts and Stick-at-Homes

Wild Animals and their Habitats

Sierra Club Books
San Francisco

Charles Scribner's Sons
New York

The Sierra Club, founded in 1892 by John Muir, has devoted itself to the study and protection of the earth's scenic and ecological resources—mountains, wetlands, woodlands, wild shores and rivers, deserts and plains. The publishing program of the Sierra Club offers books to the public as a nonprofit educational service in the hope that they may enlarge the public's understanding of the Club's basic concerns. The point of view expressed in each book, however, does not necessarily represent that of the Club. The Sierra Club has some 50 chapters coast to coast, in Canada, Hawaii, and Alaska. For information about how you may participate in its programs to preserve wilderness and the quality of life, please address inquiries to Sierra Club, 530 Bush Street, San Francisco, California 94108.

Library of Congress Cataloging in Publication Data

Milne, Lorus Johnson,
Gadabouts and stick-at-homes.

Includes index.
1. Animals, Habitations of—Juvenile literature. I. Milne, Margery Joan Greene, joint author. II. Landry, Sarah. III. Title.
QL756.M54 591.5'6 79-20743
ISBN 0-684-16473-6

This book is published simultaneously in the United States of America and in Canada—Copyright under the Berne Convention.

1 3 5 7 9 11 13 15 17 QC 20 18 16 14 12 10 8 6 4 2

Book design by Leigh McLellan

Printed in the United States of America

Contents

Green tree frog *(Hyla cinerea)*

1

How Big Is Home?

IF YOU ASK a dozen people what they think of as "home," you get many different answers. The baseball player knows it as the place where runs are scored. The airplane pilot and the bus driver regard it as a place where the family lives. The secretary and the factory worker travel more routinely every day before coming home. Generally, home is a quiet, secure place where one is least likely to be struck out by the uncontrollable hazards of life.

If you watch a dozen kinds of animals to see if they, too, have particular places to which they return for a quiet time, you find they have a great many different goals. Of course, you say, the worker honeybee comes home to the hive—the same hive—every night. She starts off on a fresh expedition each morning to seek nectar and pollen from flowers, perhaps a little resin with which to caulk a crack, and maybe a cropful of water to moisten and cool the surface of the waxy comb.

Gadabouts & Stick-at-Homes

Home is where the queen bee is, where the young bees develop, where food is stored. Home is where all the sister workers congregate each night. Home has a special odor like no other, a blend of all the various flower scents that come from the nectar and pollen the bees gather. No other hive of honeybees smells quite the same, for even the workers in the next hive have collected a somewhat different assortment of treasures from the flowers.

Many beekeepers count on the honeybees' loyalty to the hive community. They wait until nightfall, when all the bees are in the hive, then put the whole structure with its living contents on a truck and transport it through the darkness to some new location. After dawn, the bees awaken. They emerge from the hive, examine its surroundings for reliable landmarks, and set off on foraging expeditions. The insects navigate by means of a "sky compass," which is a pattern that their eyes detect (although ours cannot) in the blue sky. They adjust their direction in relation to the apparent movement of the sun from east to west by means of an inner clock. The bees depend on memory to find their way back to the hive after every trip. If their home is taken to a different location one night, the following morning the bees will learn anew their whereabouts. Almost none get lost. Day after day the same individual bees may seek nectar and pollen from orchards or fields a hundred miles or more apart. The bees pollinate the flowers they visit and ensure a crop of fruit. Their services reward the orchardist or the farmer, who pays the beekeeper to bring the hives on a visit at the proper time of the season.

Those who have a pet dog or cat see little unusual in the honeybees' loyalty to the hive. A cat or dog may tolerate being transported many miles in the family car between a familiar neighborhood and a vacation cottage—and may even feel at

Honeybee *(Apis mellifera)*

ease in a different motel room every night on an extended trip. So long as the same people, the same automobile, and, perhaps, the pet's movable bed accompany it, the animal remains unperturbed. It can venture off on short explorations and return to its human family, its customary resting place. If a cat is released at a summer camp which it has visited many times before, it may head straight for the field in which it caught a mouse on the previous trip—as though it expects another mouse to be there, waiting to be pounced on. Home for these pets centers on the people who regularly provide them with food and shelter. But if these reassurances disappear for long, the cat or dog may set off on foot, overland, for its previous territory.

The house, too, becomes home for the dog that spends many hours indoors. We learned this the first time a friend invited us to go unaccompanied to his home at night, to enter the unlocked front door and brave his large, loud-voiced dog on our way to the guest bedroom. The dog began barking with alarm as soon as it heard the unfamiliar sound of our automobile motor and car doors from the parking space outside. When we slowly opened the front door and encountered the beast face to face, the animal drew back its lips to expose its sharp teeth and protested even more vigorously. It retreated a little when we reached in to turn on the light. The dog backed up a step at a time, still barking ferociously, as we gradually entered the house, speaking to the animal by name and trying to calm its fears. Not until we had advanced through the living–dining space to the door of the guest bedroom, and the dog was in the doorway of the master bedroom, could we feel much confidence in our host's assurance that his canine guardian would not attack. Our friend's bedroom was the dog's final refuge—the place where it regularly slept at night and waited for the familiar sounds of its master's sports car as he returned. In the absence of the owner, that bedroom was the ultimate sanctuary. It alone was to be defended against any stranger.

After breakfast the following morning, our host and his docile dog accompanied us to the rocky shoreline of his property. There we found snaillike animals called "limpets" clinging to the rocks. Each of these creatures holds tightly under a low shell shaped like a coolie hat. The limpet is exposed to air whenever the tide recedes. But after the water rises and covers the limpet again, it raises its shell a little to ventilate its gills, to clear its blood of carbon dioxide in exchange for oxygen from the sea. Then the limpet begins to creep over the rock, perhaps to other rocks, to areas where a thin carpet of green plants (algae) offers nourishment. The limpet browses for a while, then starts its slow journey back home. In ten hours or less, the tide will be low again. Ordinarily the limpet will be back on its home spot, clamped there securely, before being exposed to air once more.

Along some coasts you can see and feel the homes of limpets and of other mollusks with similar habits. You discover their home sites as oval depressions in hard rock. The resident of each site may remain there, fasting, while storm waves pound upon its shell. The water rushes over the rock, up one side of the limpet shell and down the other, pressing the limpet against its support. The rock protects the rim of the shell from the rough flow, but the rim may grate against the rock each time the soft body of the animal is compressed by a wave. Scientists believe this grating action creates the oval depression in the first place and helps to deepen it. Perhaps the limpet eats all of the green plants at its home spot, permitting erosion to scour the site while it is feeding elsewhere. At the same time, the animal restores the rim of its shell, where it has been worn away by grating against the rock.

The activities of most animals are regulated by day and night, rather than by tides. Some become active only in daylight, while others wait to seek food and companionship in darkness. Two tree frogs impressed us with their emergence at nightfall and subsequent return to a resting place as the sky brightened before dawn. One of these amphibians lived near Lake Placid, Florida, and had established a home for itself within a spacious house. Its hideaway was a vase containing an attractive house plant. The vase was moved from the dining-room table to the sideboard at mealtimes, then returned to serve as a centerpiece upon the gleaming expanse of polished hardwood. After the room became dark each night, the frog climbed out of the vase, hopped to the table edge, leaped to the floor, jumped repeatedly to reach the screened wall of the adjacent porch, and fed on insects there. Before daylight, the frog made use of the adhesive pads on its long toes to climb a table leg and make its way to the vase, where it could rest without being seen.

The other tree frog revealed itself by being tardy in retiring for the day. The tiny creature, barely an inch-and-a-half long in the body, adopted for its sleeping site a birdhouse occupied at various times by house wrens, tree swallows, and chickadees. When the birds brought nest materials to the box, the frog clung to the inside wall. While the mother bird incubated her eggs or tended her young, the frog slept as best it could under the edge of the nest. Each night, while the birds dozed, the frog emerged from the entrance hole, climbed elsewhere in the tree to feed on insects, and then returned to the birdhouse before daybreak.

At dawn we often see the frog in its doorway, facing the sunrise, as though reluctant to go back inside to its sleeping site. If we approach, the frog disappears. For a night animal, our New England summer days may begin too soon and last too long. But the darkness inside the frog's shared home offers no meals, as does the brief night outside.

Home is quite another matter to the robin that builds a new nest each May on the same branch of a white pine tree close to our front porch. She returns to her chosen site as though paying no attention to her mate-of-the-year, who tries out various prominent perches high in the tree where he can keep her in view. The male watches alertly for all other birds of similar size and drives them off before they can seek a suitable nesting place in his territory. His loud caroling, begun long before dawn and often continued late into evening twilight, identifies the area specifically as robin space.

The female robin gains reassurance from her mate's song as she brings grass, weed stalks, bits of string, and beakfuls of mud. She works these materials into the form of a deep cup fitting the contours of her body. As her nest takes shape each year, she grows more protective of her home site. She dashes at any female robin that so much as perches in her vicinity, while her mate chases away any male robin. Together or separately, the pair will oust even a sparrow or a chickadee that ventures near the nest and any woodpecker or catbird that approaches the tree. Yet we, or the postman, or any quiet human visitor can walk under the nest—not a dozen feet above us—to our front door without disturbing the female in her nest or her mate caroling at the top of the pine.

A cat or a crow gets a very different reception, however. Both robins begin to call for help. *Erk! Erk! Erk!* the message goes out, often faster than once per second. Blue jays hear it and rush to the area. A gray catbird may join the fluttering group. No longer does the catbird display its amazing repertory of songs. Instead, it wails *Miauw! Miauw!* in a way we distinguish easily from the call of any hungry cat. The birds fly at the intruder and pursue it for a hundred yards or more as it moves away. Only the boldest cat or crow ignores the defenders of the home site to raid the nest for eggs or chicks. Occasionally a squirrel responds to the commotion by scrambling to the robins' branch to feast while the parents are distracted by the original intruder. The female robin must lay another set of blue eggs if she is to raise a family that year.

The size of the territory that an animal regards as home

can be judged more easily if the individual bears some distinctive marking. It may be an extra white feather, which is replaced with another like it every year, or a missing toe or torn ear due to some accident. In Africa, scientists who study giraffes or zebras or elephants now photograph both sides of each animal they encounter. From their pictures, an observer can see distinctive differences and rely on these in identifying individuals. A zebra's stripes, for example, show almost as much character as human fingerprints. Just as the skin patterns on our fingertips differ from the right hand to the left, so do the markings on the right and left sides of a zebra. Zebras use these patterns in recognizing the individuals in a herd. Their behavior shows that they can tell at a distance if an unfamiliar zebra is approaching. Scientists now realize that a young zebra or a female may be at home in one herd early in the year and in a different herd later on. Both will be accepted as newcomers almost anywhere on the same plain. A male zebra is less able to wander from herd to herd. He must stay where he is known and tolerated, unless he is either a superb fighter or extremely submissive. The dominant stallion in each herd drives off any competitors; he insists on being the center of attention. Home for "his" herd changes according to where he goes.

In the trackless oceans, groups of whales—chiefly females—travel together for years. Whalers call these groups "pods." Long before anyone found a way to confirm the idea, whalers insisted that male sperm whales course through the oceans of the world on individual schedules according to the calendar. (Old whales, recognized by markings and scars, were seen by whalers year after year.) Supposedly, each male arrives regularly at each place along his course through the Atlantic and Pacific oceans and far to the south around Antarctica. He has mates along the way, females that do not venture so far. For these male sperm whales, all of the seas together comprise what we think of as home. Male and female whales have differing views of how big home should be.

Atlantic salmon *(Salmo salar)*

2

Animals with Identification Marks

When he was a young man on his father's farm in Pennsylvania, John James Audubon noticed five little phoebes in their parents' nest along Perkiomen Creek. He wondered whether the nestlings ever would return to the same location. To satisfy his curiosity, he made five lightweight bracelets from silver wire, which would not rust. He fastened the bracelets loosely around the ankles of the nestlings and made sure that these markers would neither fall off nor hurt the phoebes in any way. The following summer Audubon was delighted to see two of the five again. Each had reached maturity with its silver anklet attached and was busy at a nest, raising a family of its own along the Perkiomen.

Audubon never realized how fortunate he was in finding any from such a small number of marked birds. With the best of modern methods, it is unusual for more than a small percentage of tagged creatures to be observed a second time.

The idea of using metal tags with distinctive numbers came at the end of the nineteenth century. It appealed independently to a fisheries officer in the United States and an ornithologist in Denmark. Charles G. Atkinds began clipping numbered tags to the fins of Atlantic salmon, which he caught and released in the Penobscot River at Bucksport, Maine. The fish swam upstream, left fertilized eggs to develop in fresh water, and returned to the sea. Many came back to the Penobscot to repeat the process in subsequent years. Fishermen and cannery owners helped Atkinds recover a fair number of the hundreds of tags he had attached to salmon. From the tag numbers, as recorded in his notebook, Atkinds learned that few fish stray to other rivers. Almost all return to the same stream in which they swam in previous years, the one in which they originally hatched from an egg. To them that particular stream is home, even after years at sea and many miles of travel.

A Danish schoolmaster, H. C. C. Mortensen, cut and smoothed small, flat strips of rustproof metal and scratched a number on each one. He curved these strips into anklets around the lower legs of birds he caught. Mortensen said he was "ringing" the individual, giving it an identity with a numbered ring. He made rings of different sizes and put them on starlings, storks, small ducks known as teals, and a few birds of prey. His success in learning how far these marked birds flew before being caught again, how long they lived, and what happened to them drew the attention of ornithologists throughout Europe.

An American biologist brought news of Mortensen's method to the United States. Soon hundreds of people were participating in attaching numbered bands to millions of birds. A central records file was established. The file is now handled by computer at the Patuxent Research Center of the United States Fish and Wildlife Service in Laurel, Maryland, and includes data for all birds that are banded in the United States and Canada. Information on the migration route of each kind of flier gradually fits together into meaningful patterns and can prove invaluable in efforts to preserve bird species, particularly those that are hunted. For example, a duck or goose that escapes being shot early in its migration becomes a possible target in the next state along its route. Hunting laws must be adjusted in each state to make sure that plenty of birds get through unharmed, to maintain the population. Lawmakers need to know how many of each species travel between Canada and the southern United States, how many are shot, and where the survivors go. These facts can be learned by recovery of the numbered bands on birds that are shot or caught.

North American waterfowl migrate regularly between the place where each bird is hatched and wherever it spends the winter. Four major routes, called "flyways," have been identified. The most popular, called the Mississippi Flyway, runs through the center of the continent to marshes along the Gulf of Mexico for the cold months, then northward into central Canada each spring, ranging as far west as the Mackenzie River and as far east as Greenland. The Atlantic Flyway, which lies near or along the Atlantic coast, is used by some birds from Greenland and the eastern half of Canada and runs through Florida to Cuba. Some of these fliers then head straight across the Gulf of Mexico to Colombia, while others follow the chain of islands along the eastern edge of the Caribbean Sea to Venezuela and southward. A third route, known as the Central Flyway, lies along the western edge of the great Mississippi River basin. The primary users of this route are land birds from the Rocky Mountain region and waterfowl from the Great Plains. Ducks from the Mackenzie River delta and northern Alaska use this flyway, but few venture farther south than the Rio Grande or the Mexican state of Veracruz. The fourth flyway, called the Pacific Flyway, runs along the Pacific coast. Although it serves most of Alaska and the Yukon, western Canada, and the states on the western side of the Rocky Mountains, it is the least traveled of the four major routes. Birds that are hatched in these regions travel shorter distances because the weather is so uniform. A few users, such as hummingbirds, proceed south beyond the forests of Guatemala.

Information gained through recovering the numbered bands of migrating birds shows that they travel with others of their species from their nesting grounds. Young individuals learn the way from older birds; yet many a bird can navigate on its own, without guidance from more experienced fliers. The existence of this mysterious ability was proven one year by Professor William Rowan. The professor kept fifty-four young crows captive in a large pen near Edmonton, Alberta, Canada. No adult crows were allowed close, nor could the young birds see the mature crows when they began flying away for the winter.

Rowan released the fifty-four juveniles a full month after the last resident crow had been seen flying south. By then the prairies were covered with snow, and the temperature had dropped to near zero degrees Fahrenheit. Off went the young crows, free to travel in any direction. Virtually every one of these birds flew southeast. Within two weeks more than half of them had been caught with food for bait. Each was examined and freed again. One had journeyed 250 miles. All were headed along a narrow pathway between Edmonton and central Oklahoma, which is the wintering ground for almost every Alberta crow.

Fourteen sizes of aluminum bands currently are employed to permit later recognition of birds ranging from small songbirds to swans, pelicans, and eagles. The smallest size can be fitted to a hummingbird. These bands last for many years on birds that do not visit salt water often. Bracelets made from stainless steel, which sea water does not weaken, are needed for sea birds, loons, and other fliers that spend the winter months along the coast. The United States Fish and Wildlife Service supplies the banders at cost; each band is imprinted with the lettering "Write F & W Serv Wash D.C. USA" and a serial number. Anyone who finds a banded bird can help by reporting the number and the date and place of recovery to the Service, whether the animal is alive or not. The number is checked against the central file, and the new information is added to the records. The Service thanks the finder by informing the person of the history on file for the banded individual: the species and sex of the bird or bat and its probable age; who attached the marker, where and when; and any facts about where it has been sighted before.

Scientists learn the behavior of the various species by studying the records of banded birds that have been recovered. Because there is information on file regarding hundreds or thousands of individuals of each species, it is easy to plot on a map where a species goes to reproduce, what routes it follows to wintering grounds, and how large or small an area is needed as a refuge from northern cold. All of the Kirtland's warblers, for example, nest among young jack pines in the woodlands of northern Michigan, then fly to the Bahamas for the winter. The chimney swifts from all over North America journey to a few small valleys in the mountains of Peru, where they feed on insects until it is time to fly north again. Some chickadees come to bird feeders in New England during the winter, then travel far into Canada to nest. Other chickadees, identical in appearance (except for their numbered bands), spend the cold months in the southern Appalachian Mountains, then migrate to New England in the summer to raise a family and visit feeders. Even the number of each species can be estimated from facts revealed by recovered bands.

Wild mammals can be marked, too, though it isn't wise to band them with a metal bracelet, since it can get caught and cause trouble. An animal the size of a rabbit or a raccoon gets its feet into too many tight places to wear this type of band. Far better are flat markers made of stainless steel and bent into a V. They come in pairs imprinted with the same number and are closed permanently over the edge of each ear. If the mammal loses one ear and its tag in a fight, the second tag will remain in place. A deer or an antelope sometimes is given a conspicuous plastic collar, fastened loosely around its neck. The collar bears large numbers or a combination of letters. It can be seen from a distance through field glasses, even from a low-flying airplane, and identifies the animal as surely as the number and colors on a football player's uniform.

Mice, squirrels, and other small mammals seldom live long enough to warrant giving them metal tags. Simpler ways can be used to identify these animals for a month or more. At the University of California in Berkeley, Dr. O. P. Pearson rebuilt ordinary mouse traps to serve this purpose. He cut and bent the spring mechanism so that it would not hurt the animal. Instead, it struck the rubber bulb of an eyedropper full of colored dye. If a mouse sprung the trap, dye was squirted onto its tail end. Each trap held dye of a different color. Any mouse that visited several traps soon wore a coat of many colors. Whenever Dr. Pearson caught a marked mouse in one of his harmless traps, he could tell where the animal had been and how large a territory it had explored. He also shaved a small area on the back of some mice and tattooed the skin there with a number he could recognize again. He soon learned how many mice lived near his traps and how long they kept coming before being replaced by other individuals.

Since 1926, people with an interest in whales have been marking these huge animals far at sea. Only in this way can anyone learn how many there are of each species of whale, how rapidly they grow, and where and when they travel. Whalers do the marking. From a small, fast boat they shoot a stainless-steel dart into the fat below the skin on the back of the whale. Usually this is done on mother whales and babies that accompany them, since neither legally may be caught. The dart bears the inscription "Reward for return to Admiralty, London" and a number. The whalers record the date, the place, and the kind of whale that has been marked and supply this information to the British Admiralty Office in London, England. The dart is recovered later, elsewhere, if the whale is caught. The whalers exchange the information about place and date of recovery for the reward promised on the dart. Scientists associated with the Admiralty study all the facts and keep members of the whaling industry informed. Facts on ages and travels are used to replace guesswork in determining regulations for the benefit of whales.

On land, large and dangerous animals have to be handled with special care if they are to be given identifying marks. Generally, a game warden carries a special gun that shoots darts, instead of bullets, when he stalks a wolf, bear, tiger, or elephant. When the dart hits an animal, usually as it is running away, a charge of carbon dioxide gas explodes; this drives a hypodermic needle into the flesh, releasing a measured quantity of a tranquilizing drug. In a few minutes the animal loses its ability to run away, or even to stand up. The game warden and his assistants hurry to wherever the helpless animal lies, examine it quickly, perhaps weigh it, and attach ear tags or other markers. Then they inject an antidote, which neutralizes the action of the tranquilizer. By the time the game warden and his crew have retreated a safe distance, the animal is able to get to its feet and go off on its own business.

Often the marking device for a large animal consists of a harness containing a tiny radio transmitter and long-lasting batteries. So long as the batteries supply energy, the radio sends out a message once each minute, day and night. Anyone with suitable equipment can pick up the message and learn the direction from which it comes. The individual animal can be located within a few yards by a person on foot, in an off-road vehicle such as a jeep, or in a small airplane.

By tracking the radio signals every day, the travels of the animal can be plotted on a map. The movements of elephants equipped with particularly strong radio transmitters can be followed by satellite for months, over hundreds of miles. The radio transmitter reports the animal's location every time the satellite goes over — sometimes twice a day. The information is relayed back to earth as the satellite overflies its base station. The scientist who collects this information can follow the animal's progress week by week and gain a good idea of what the animal is doing. It makes no difference that the elephant with the radio transmitter is south of the equator in Africa while the scientist is in New York or London. He or she can keep track of it. In fact, the scientist can gather reports on many different animals that the satellite passes, so long as each animal has been equipped to broadcast its position on its own particular wavelength of radio transmission.

Even insects can be marked. One friend showed us how he was making individual bumblebees identifiable by dabbing a small dot of paint on the bare back of each bee as it visited a flower in his garden. By placing some dots forward on the right or left and others farther back on one side or the other, and by varying the color of paint, he produced distinctive patterns he could match against a code chart. He looked for the marked bees in other gardens and learned much about the territories that individual bees explored for food.

Dr. Frederick A. Urquhart and his wife, Norah, invented a way to mark thousands of butterflies. The Urquharts catch the insects while they are still asleep before dawn. Their most fascinating study involved marking monarch butterflies, which migrate in late September. These insects fly southward sociably, from Canada and northern states, and avoid harsh northern winters. Along the way they cling in great clusters as they sleep at night, particularly from oak trees. Large numbers can be caught, but they must be handled individually with great care.

To mark the butterflies, a small area near the leading edge of one front wing is rubbed free of microscopic scales on both its top and bottom sides. This exposes a surface to which the marker will stick. The tag itself is a small rectangle of waterproof paper, printed on one side and self-adhesive on the other. It is folded around the edge of the wing and pressed firmly against the bare areas top and bottom. The Urquharts flight-test the tagged insect inside a screened porch before freeing it to travel. The butterfly weighs forty times as much as its tag but must not be hampered by it. If the tag is applied correctly, the insect flaps along normally, unaware of the white marker with the printed words "Send to Museum, Toronto, Canada," and a number. The Canadian Postal Service routes any message from the finder of a monarch to the Royal Ontario Museum of Zoology and to Dr. Urquhart. The message cites the tag number and the date and place where the insect was seen or caught.

Monarch butterflies *(Danaus plexippus)*

For a while, four monarchs tagged on September 18, 1957, held the record for travel. All flew across Lake Ontario from Highland Creek on the north shore near Toronto. One was caught on October 5 in Roxie, Mississippi — 1060 miles away. A second reached Atlanta, Georgia, on October 9—a mere 740 miles, yet a long distance for a butterfly. A third was sighted 985 miles away in Niceville, Florida, on October 11. The fourth turned up on January 25 in San Luis Potosí, Mexico—1870 miles from where it had received its distinctive marker.

Of all these studies with marked monarchs, the record breaker came on December 18, 1976. Two collaborators of the Urquharts, Ken and Cathy Brugger, discovered thousands of monarch butterflies wintering together on a mountaintop in Mexico. Many wore wing tags. Hundreds of others were given similar markers in the next few days. The first among several winter destinations of millions of monarchs had been discovered. Some of these insects had come from Wisconsin, others from Nova Scotia. Several were seen again later in the Gulf states as they traveled partway north at the beginning of spring.

An identifying mark helps us learn about the creatures with which we share the earth. It can answer such questions as whether or not we have met the individual before. If we have, it can tell us where. It also can show us how long this wild creature has lived since the marker was attached, how far it has traveled, and something about its traveling schedule. Finally, the marker lets us know whether the animal is a gadabout or a stick-at-home.

Arctic terns *(Sterna paradisaea)*

3

The Greatest Traveler of All

AS THE NIGHTS lengthen and the noonday sun swings lower in the sky, the American robin no longer can find earthworms on the ground at dawn or following a cold rain. The worms have moved deep into the soil to avoid the frost of coming months. Already the robins have eaten most of the conspicuous fruits that appeal to them, leaving few exposed where snow will not conceal them until spring. These scant rations may permit an occasional robin to survive the winter, but the majority fly south, where the weather will be less severe. Robins travel in flocks, young and old together, moving a little farther south with each autumn storm or when they have largely exhausted the local food supply. In mild winters the birds may not migrate at all. Severe conditions impel them to Florida and other states bordering the Gulf of Mexico. Even there they seem uneasy, impatient to regroup and fly north again. Sometimes a flock returns to the Northeast the day before a late-season snowstorm, encountering conditions that must be hazardous indeed.

Whooping cranes migrate seasonally, too, and their travels are very closely watched; these stately birds almost vanished from the earth a few decades ago. Today, slowly increasing numbers of whoopers flap across the heartland of the continent like flying crosses, their necks outstretched, wings spread seven-and-a-half feet across, and long legs trailing. They shuttle between their nesting grounds near Great Slave Lake in northern Canada and their winter refuge along the Texas coast. Game wardens and bird watchers excitedly report sighting small family groups along the 2300-mile migration route. Hunters are forbidden to shoot them, so that the whooper population can continue upward from its alarming low of fourteen individuals in 1939.

These handsome white birds, fifty inches tall with little red caps and long beaks, used to breed in Iowa and other states from coast to coast. But after the end of the nineteenth century, the only survivors were those that flew northward each spring into Canada, to some unknown destination. No one knew the secret of where they nested until the summer of 1956. Then, from a helicopter in a remote part of Wood Buffalo National Park, several mated pairs were sighted, each tending a gangling chick or two. Luckily the surviving cranes followed an ancestral pattern of nesting in one of the safest regions available to them in the whole of North America; it had been set aside for Canada's dwindling herd of native bison and was closed to settlers. The cranes have no trouble traveling the 2500 miles to their winter homes in Texas and then covering the same distance back to Canada.

Distance means even less to other kinds of birds that nest

still farther north, beyond the Arctic Circle. They arrive in unbelievable numbers on the tundra, raise their families, and depart again without delay. In the long days of polar summer, food becomes extraordinarily abundant in and around melt-water pools. The birds take full advantage of this in feeding their chicks and themselves; they scarcely take time to sleep. Only the rate at which the young birds can digest their meals seems to limit their speed of development. Many are ready to fly south before the earliest signs of harsh autumn weather, after a summer less than three months long.

One of these migrants remained a mystery until June 12, 1948. The bird was known to arrive in Alaska from across the broad Pacific Ocean each summer, but then it would disappear. This traveler, "no bigger than a pullet, but strong of wing," winters in Tahiti and elsewhere far to the southwest of Hawaii. Known as the bristle-thighed curlew, it was the only bird of 815 species known to breed in the Americas north of Mexico for which no person had seen a nest or chick or had identified the summer home. Even when it touched down in Hawaii on its travels between the South Pacific and Alaska, it had at least 2800 miles to fly over open sea between one landing site and the next. Like other birds that nest so far north, it migrated farther in both directions than anyone thought reasonable.

The curlew's secret was discovered by an exploration team led by Dr. Arthur A. Allen of Cornell University, with the aid of two local schoolteachers in westernmost Alaska. The first nests of the bristle-thighed curlew to be located were merely slight depressions in the ground among low clumps of black lichens and inconspicuous shrubby growths of arctic-alpine azalea. These two plants grow slowly almost everywhere in the Far

North and on mountain peaks farther from the pole in both North America and Eurasia. The bristle-thighed curlews find this combination most attractive close to the mouth of the Yukon River. The birds fly thousands of miles to reach the place and later retrace their route in the opposite direction.

Dr. Allen found the nests of many other kinds of birds where the mysterious curlews were found. These other migrants travel by various paths to reach the same small area of tundra. Among the curlews' neighbors are little shorebirds the size and shape of killdeers, including golden plovers and ruddy turnstones. The Alaskan plovers long have been known to fly straight south over 2800 miles of Pacific Ocean to Hawaii and Wake Island, then southwest an equal or greater distance to Australia, New Zealand, and such isolated islands as the Tuamotu (the "Low Archipelago"), where they again may have bristle-thighed curlews as neighbors. But the plovers feed along the shore, while the curlews often raid the nests of birds native to the Southern Hemisphere. Curlews steal food from southern terns, frigate birds, boobies, and albatrosses.

A separate, larger population of golden plovers nests on polar tundras all around the Arctic Ocean. These amazing birds fly to Labrador, down the Atlantic coast a short distance, out over open water to the West Indies, and then onward to Argentina. On their return trip early in the northern spring, those birds native to North America take a more westward flyway. They cross the vast Amazon basin and the Gulf of Mexico, then travel up the Mississippi River basin and over the prairie provinces of Canada. By following this annual loop, they find food at rest stops along the way. At each stop they fatten up for the next stage of their tremendous journey. Biologists from the University of Florida discovered on Wake

Island that golden plovers accumulate as much as forty grams of fat before departure. It is enough fuel to permit them to fly up to 5800 miles nonstop.

Ruddy turnstones are far less ambitious travelers. After they raise their young on the northern tundras, they fly south to either lower California or Florida for the winter. We see them there in January, basking in the warm sun or scavenging for sand hoppers and other small animals along the wet parts of the beach. As though enjoying a lazy life, they wait for the nights to shorten and winter to free its grip on polar lands.

All across the northern continents, much more energetic birds seek out the same tundras at nesting season. Their slim bodies are pale gray, their forked tails pure white. They blend inconspicuously with dirty snow early in the season as they settle on short legs. Each bird's legs, as well as its feet and beak, are blood-red. The black eyes scarcely show because a cap of black feathers comes down, like a shadow, in line with them along both sides of the head. These birds are well-named the arctic terns, for they follow the spring toward the North Pole as far as land extends. In North America they rarely nest farther south than northern British Columbia or, along the Atlantic coast, beyond the islands off the shores of Maine. Occasionally a few arctic terns breed on lonely Muskegat Island, located between Nantucket and Martha's Vineyard near Cape Cod.

The nest of an arctic tern often is no more than a saucer-shaped hollow in a rock or a shallow cleft that can accommodate two eggs. It may be a depression in sand or in a covering of moss. Rarely does a parent bird contribute any extra material, such as fine grasses, to prevent its eggs from rolling. The eggs hatch after three weeks, and fuzzy little spotted chicks step forth. Generally the youngsters crouch as inconspicuously as they can unless their parents are there, feeding them.

Adult terns return frequently to their chicks with partly digested fish, which they catch by diving shallowly in nearby cold sea waters. The young birds benefit, too, from a sort of flying umbrella of parents, who hover above the nest sites and drive away marauding gulls. Seldom are the eggs or chicks of one pair of terns less than a yard from the nearest neighbor;

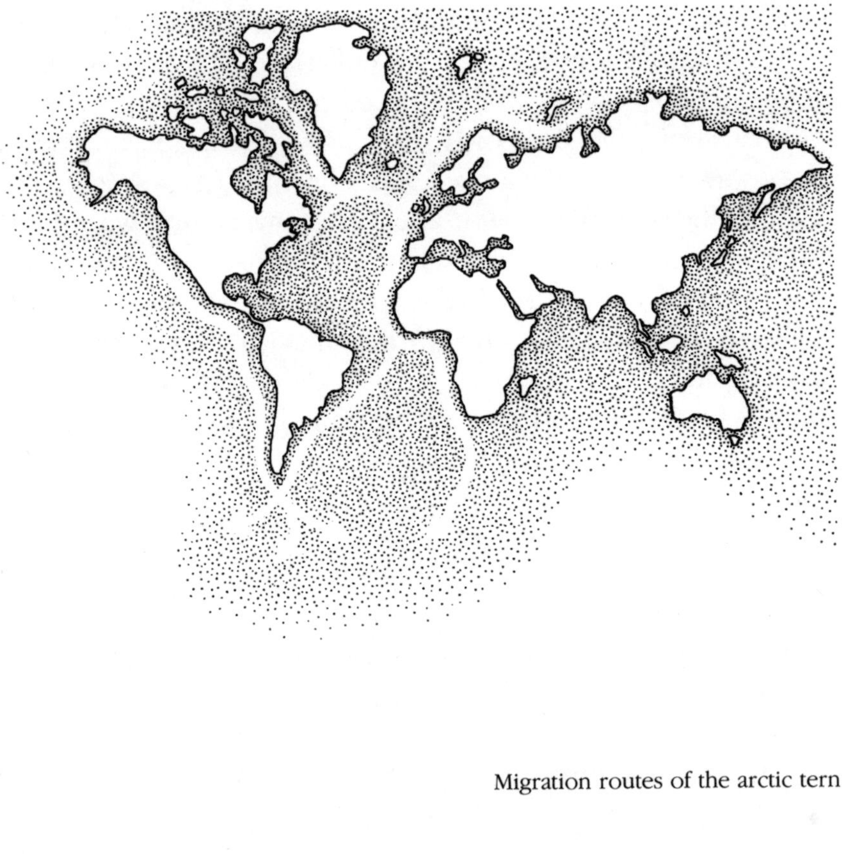

Migration routes of the arctic tern

yet these birds succeed in rearing their young more often when the colony is large than when only a few pairs have nested in modest proximity. The shared care for the next generation continues until, after about fourteen weeks, the juveniles are able to fly on their own. Then all of the arctic terns depart, in as much of a group as when they arrived at the beginning of their nesting season.

From the polar coasts of North America, the arctic terns join those from New England in an eastward sweep across the North Atlantic. Near the British Isles they combine flocks with terns from the Eurasian northland and wing past France, Portugal, and the westward bulge of Africa. Off the shores of Guinea and Senegal, the birds separate into two great streams. One follows the African shoreline to its southern tip before veering southwest toward Antarctica. The other stream crosses the South Atlantic to the coast of Argentina, then continues southward to Tierra del Fuego and the Falkland Islands. There the birds encounter arctic terns from southern Alaska and adjacent Siberia, fliers that have directed their course along the Pacific coast of both Americas. Thence all of the terns fan out over sub-Antarctic waters, rarely in sight of the ice-clad continent of Antarctica. They spend the southern summer between the scattered islands, feeding on small fishes and the abundant shrimplike crustaceans known as krill.

The tremendous journeys of the arctic terns cover a minimum distance of 11,000 miles. Since the birds pause and wheel at intervals to feed along the way, their actual course is circuitous. Many of them must fly more than 25,000 miles each year, not counting the shuttle service they maintain to carry food to their hungry chicks.

Arctic terns easily hold their place as the world's champion migrants. In consequence, as W. W. Cooke pointed out, they "have more hours of daylight and sunlight than any other animal on the globe. At their most northern nesting site the midnight sun has already appeared before their arrival, and it never sets during their entire stay at the breeding grounds. During two months of their sojourn in the Antarctic they do not see a sunset, and for the rest of the time the sun dips only a little way below the horizon and broad daylight continues all night. The birds, therefore, have twenty-four hours of daylight for at least eight months in the year, and during the other four months have considerably more daylight than darkness."

A Swedish oceanographer, Bertil Kullenberg, marveled that these terns fly so circuitously over the seas, rather than head directly for their destinations, since they have so many miles to go. Then he noticed how scrupulously the birds avoid warm water. They will detour for thousands of miles to fly over cold currents, which often correspond to regions where deep water rises to the surface, enriching the production of drifting plants and thus of food fishes and krill. In many instances, the sky paths of the terns follow the green "veins" in which the whalebone whales formerly migrated in great herds. Now that whalers have so greatly reduced the whale population, the terns may find less competition when seeking krill to eat.

For nearly ten months each year, arctic terns remain far from land, looking for small creatures of the sea that can be caught close to the ocean's surface. Long, beating wings and an expressive tail keep the tern in motion, its few ounces of body well nourished but gracefully slim. The bird follows its inherited schedule all the way south and then back north again, sometimes to find its familiar nest site hidden by a late-season snowstorm. Digging down through the white blanket, this greatest traveler of all prepares a place to lay two eggs and incubate them. Even after so long a journey, the bird takes almost no time to rest. It soon becomes busy caring for the next generation of global fliers.

Chimney swifts *(Chaetura pelagica)*

4

The Chimney Swift's Two Homes

NORTH AMERICAN INDIANS had many names for the small, gray bird that roosts every night and nests each summer inside hollow trees. Yet the bird attracted little notice until European colonists arrived. The settlers built houses, churches, schools, and factories, all with chimneys as part of their winter heating system. The birds discovered that many chimneys were left unused all summer. They were perfect substitutes for hollow trees and could be used again year after year. Quietly the birds began moving close to people. They became known as chimney swifts, and their numbers increased as people built more chimneys.

A more improbable bird would be hard to find. The chimney swift also has been called a "flying cigar." Its slim body, no more than five-and-a-half inches long, tapers to a point at the beak end and seems to have no expansion at the tail. The bird appears to beat its wings alternately instead of simultaneously, although high-speed motion pictures reveal that this is an illusion. A chimney swift often rolls from side to side so rapidly while in flight that we see its right wing in silhouette where its left wing was a moment earlier. Actually, the bird flutters its narrow wings instead of flapping them or it holds them motionless while soaring in great spirals and circles. In flight, its wings are outstretched in a curve as smooth as a parenthesis.

The chimney swift is a supreme aerialist, never pausing to perch anywhere. It holds its short legs and strong claws so tightly compressed to the contour of its feathers that it gives the appearance of having no legs at all. Scientists chose the Greek words for "the legless ones" in naming the bird family to which the swifts belong, the *Apodidae.* The only times a swift shows its feet are when it is clutching a broken twig in its claws or while it is clinging to some vertical surface.

The bird obtains its nest material by flying deliberately against the projecting tip of a dead branch and breaking off a piece about an inch or two in length. The trophy usually is between one-sixteenth and one-eighth of an inch in diameter and will be glued together with about a hundred others to create a nesting shelf in some hollow tree, chimney, cistern, well, or even an outhouse. Only at the last moment must the chimney swift transfer the twig from its claws to its beak, so as to have its feet free to cling and clamber as it alights inside its refuge. The glue for nest construction is the bird's own extraordinarily viscous saliva, which acts like contact cement.

Occasionally a nest with young swifts breaks off and falls, or an adult swift loses its grip on the inside of a chimney, so that a bird accidentally drops into a room where people are. No feathered creature seems gentler or less fearful. If the bird is picked up, it snuggles into a human hand. It will cling to a sweater or a jacket and may even go to sleep there if left undisturbed. It makes no attempt to bite or defend itself and appears reluctant to fly away when taken outdoors into the sunshine and given an opportunity to use its wings. This is quite normal if the bird is very young, for each nestling needs almost four weeks of parental care before being able to fly on its own. The parents come back from flying in pursuit of insects and regurgitate the food for the young birds. Eventually, each youngster is supposed to climb up to daylight and take off. Sometimes the birds can be heard exercising their wings inside a chimney, getting practice while still clinging to a support.

The insects that chimney swifts catch while flying are all small and tend to be those that are most active by day. They include bark beetles—which do great damage to trees by starting families of grublike larvae under the bark—and flying ants. Swifts also eat plant bugs and flying aphids, which suck juices from innumerable plants and transmit plant diseases. If the insect is small enough for the chimney swift to catch in its gaping mouth, the bird will pursue it and snap it from the air. No house fly is fast enough to escape. Since so many of the creatures eaten by chimney swifts cause harm to crops or to people more directly, we regard the swift as a helpful bird, a worker in our favor.

Throughout the United States and southern Canada east of the Rocky Mountains, people have grown used to noting the date when chimney swifts arrive in the spring. It is a happy event of mid-March in New Orleans, of early April in northern Florida and Philadelphia, of mid-April in Boston and Chicago, of early May in Portland, Maine, and Fargo, North Dakota. The welcome summer guests fly almost continuously during the lengthening hours of daylight, chittering pleasantly in excited

Young chimney swifts

cries that attract the attention of bird watchers at ground level. A good supply of flying insects in the air high above field and forest, town and country, seems to be the bait that lures these birds northward with the advance of spring.

At first the chimney swifts perform their aerial acrobatics mostly as solo performers. They are filling their stomachs and renewing the energy they spent in migrating northward. But in less than two weeks the fliers begin chasing one another in twosomes, threesomes, foursomes, and larger groups. Their chipping calls become sharper, louder, and repeated more often. Each noisy race loses some participants until only one pair remains, recklessly rushing on, their wings almost touching. Suddenly the two soar on at their dizzy pace, one bird directly above the other. They mate in mid-flight, with just a moment of contact. For any other kind of bird, this would seem an impossible maneuver. But for a chimney swift, which can slake its thirst by snapping up a mouthful of water from a pond without slowing its flight perceptibly, the feat is routine. Long practice on the wing makes it perfect.

The biggest danger for a chimney swift is not a hawk, which the small bird usually can outfly, but a week of rain that washes all insects from the air. The swifts can starve for lack of any way to get nourishment other than by pursuit. Cold weather, which keeps insects too chilled to fly on cloudy days, has the same effect. Less food and shortening days combine to start chimney swifts on their way south by late August from Manitoba, by mid-September from Quebec City and Nova Scotia, and before the end of October from even the southernmost United States. On their way, great flocks of the birds often are seen funneling downward like a tornado cloud into some tall factory chimney at sunset. Experience may help the swifts determine which ones retain warmth from daytime industrial furnaces but emit no suffocating gases before dawn.

For many years ornithologists, who want to learn all they can about every species of bird, could not understand why the winter home of chimney swifts remained a complete mystery. How could so common a flier simply vanish each November and reappear in March? Nearly half a million chimney swifts were caught and banded with numbered aluminum anklets in hope of learning the answer. Not a single band was recovered from any of these marked birds outside the United States. Solitary chimney swifts or small flocks of them occasionally were reported in tropical locations: the West Indies, Mexico, Central America, Panama, even near the city of Manaus on the Amazon River in Brazil. No one could be sure that these birds had come from North America or that they were not lost. One serious-minded ornithologist wondered if the time had come for bird watchers to start looking for chimney swifts hibernating somewhere in the southern states. Perhaps the stories about swallows invented in the Middle Ages applied to this improbable bird. It was said that flocks of them dove into the cold mud of southern swamps and held their breath all winter. Too much was known about the upper atmosphere to believe that the birds—any birds—flew to the moon and disappeared a while from human view.

Early in 1944 the mystery began to come clear, although no ornithologist realized this. A group of Indians who lived in a wild part of eastern Peru brought a present to some missionaries who had come to work in the vicinity. Their gift was thirteen small pieces of gray metal, each indented with magic markings. The missionaries' inability to communicate with the Indians in a common language kept them from learning exactly where the bands had been obtained. Gestures suggested that small birds had something to do with the discovery. In fact, the Indians had killed thirteen banded swifts and had kept their aluminum anklets. The place was somewhere in the Yanayaco Valley almost due east of the capital city of Lima, not far from the Brazilian frontier.

A few months later one of the missionaries journeyed to Lima to get more supplies. He took the metal bands along and turned them over to the United States ambassador. The two men agreed that the marks on the inner surface of the bands probably were abbreviations in English. They seemed to read "Return to Wash. D.C. USA." Outside, each band bore Arabic numerals. The ambassador wrote a report and sent it with the bands in a diplomatic mail pouch from the embassy to Washington, D.C. Dated May 23, 1944, the report stated that the numbered items had been recovered "from some birds killed by Indians on the River Yanayaco in the region between the Putumayo and the Napo Rivers." The missionary had said that the Indians probably got the bands about six months earlier, which would have been in late November or early December 1943. The best maps showed that the area was close to the western edge of the great Amazon River basin.

Eventually the bands reached the Patuxent Research Center of the United States Fish and Wildlife Service in Laurel, Maryland. A technician looked up the thirteen numbers in the master file and rushed excitedly with the news to the migration specialist, Frederick C. Lincoln. At last the winter home of chimney swifts had been found. And not from just one banded bird—from thirteen of them! Five bands came from Memphis, Tennessee, and had been attached by Ben B. Coffey, Jr. Three more were from Nashville. The oldest, from Opelika, Alabama, dated from October 1936. One each came from Macon, Georgia; New Haven, Connecticut; Lake Forest, Illinois; and Kingston, Ontario, Canada.

Coffey and the other bird banders whose tags had been recovered so miraculously received notification. They got in touch with one another and joyously celebrated the end of the mystery. Coffey checked through his records and found that he had attached bands to more than 35,000 chimney swifts; his five recoveries dated from September and October of 1937, 1938, 1939, and 1940. Mrs. F. C. Laskey of Nashville learned from her notebook that the one bird with her number on it was from a flock of more than 5000 that she and four other workers had banded on October 13, 1940. Between 1937 and 1942, this small group of people had banded more than 43,500 chimney swifts in Nashville, often laboring at the task from dawn until dusk.

One of the Nashville group, who had not been lucky enough to have a numbered band among the three returned from that city, was Albert Ganier. He served as editor of a quarterly magazine called *Migrant* and agreed that the next issue, which would be number three of volume fifteen, should be a special chimney-swift issue in celebration of the great event. The October 1944 issue of *The Auk,* published by the prestigious American Ornithologists' Union, included a review by Frederick C. Lincoln of all the information on file at the United States Fish and Wildlife Service. "Chimney Swift's Winter Home Discovered" announced the headline to bird enthusiasts all over the world.

The Yanayaco Valley of Peru remains wild and remote, its Indians still reserved and primitive. But the Indians have contributed no more magic bands from chimney swifts. Apparently the birds are hard to catch, hiding through our winter months in hollow trees that have no opening at the bottom. The tall trees, mostly dead and widely scattered throughout the rain forest, project up through the forest canopy. The swifts pursue insects above the trees all day, out of sight of people. As the sun goes down they seek out their shelters, where they await the following dawn. So far as anyone knows, all of the chimney swifts of North America fly to the same winter home.

These improbable birds are commuters on a yearly, rather than a daily, schedule. From November until the end of February, they are at home in a tropical country in one of the least civilized parts of the Americas. From March through October, they work above the roofs, streets, public parks, and dense suburbs of the North Temperate Zone. They overfly the chimneys, catching whatever insects travel high above the ground in summer months. Even a rain shower presents an opportunity—a chance to take a bath while flying along at many miles per hour. The birds can preen in flight just as well as they can manage most of the other tasks for which feathered creatures ordinarily perch or rest.

Only when a chimney swift is about to settle inside some dark cavity that is open to the sky does it brake its rapid flight. Suddenly it spreads its tail feathers and pulls them downward like the landing flaps on an airplane wing. So expert is the flier that it can adjust its speed to back vertically down a chimney. Out come its legs and claws to grip the inside wall; the tips of its tail feathers have projecting spines that dig in as extra support. The swift is home—at one of its two homes. That the two may be five thousand miles apart, and so different, seems not to matter in the least.

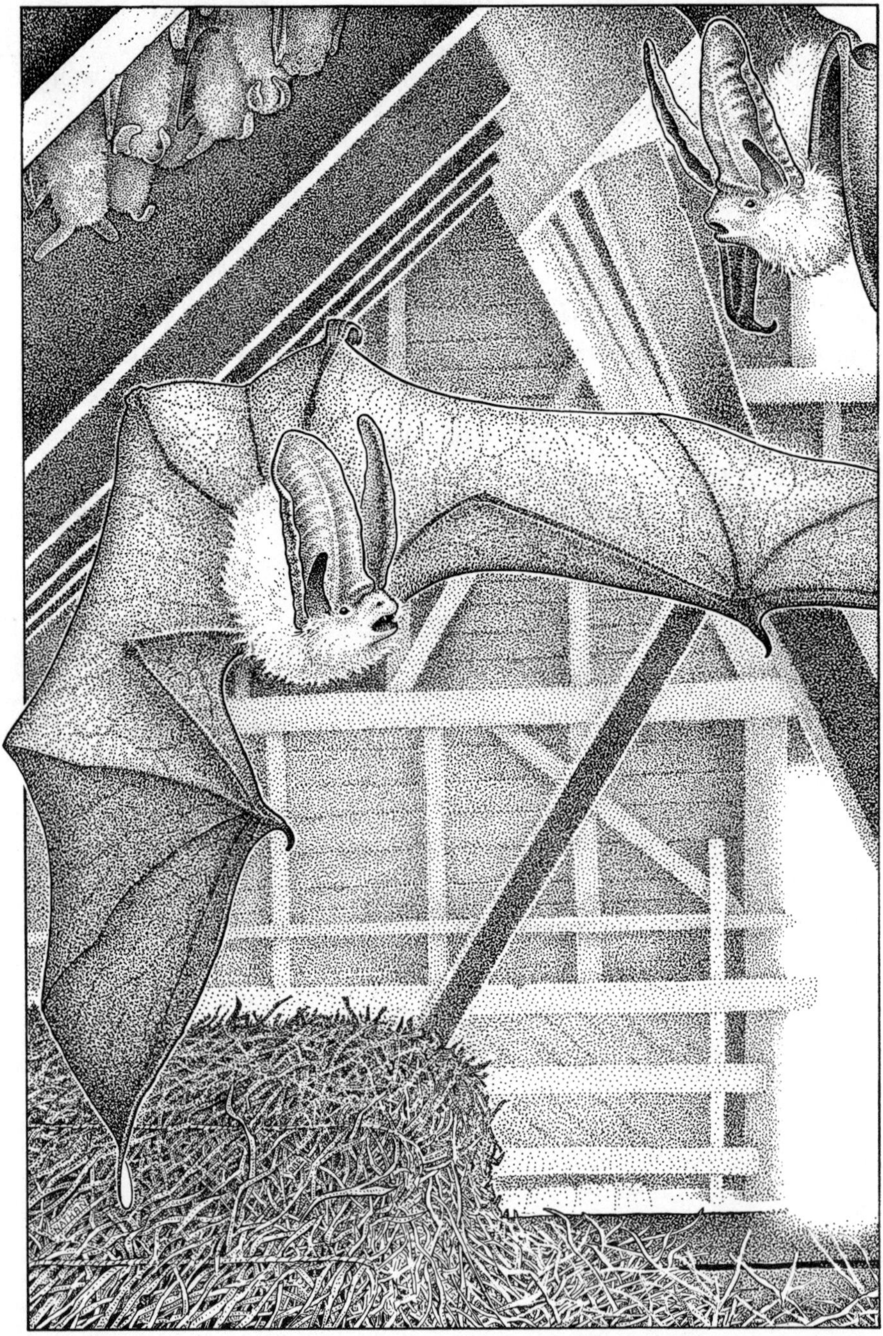

Western big-eared bat *(Plecotus townsendi)*

5

Bats and Their Hideaways

THE BARN WAS fairly new, compared with barns that people use in Pennsylvania and Vermont and Ohio; yet in California it was almost an antique. A farmer had built the barn of inexpensive wood about thirty years ago, with help from his family, friends, and neighbors. For a while the barn had been home to several cows and more than a dozen chickens, but later it came to be used merely as storage space for the retired couple who had bought the farm, barn and house and all. No cows, sheep, or chickens lived on the farm. Not even a dog or cat patrolled the property.

The continuous quiet in the barn suited a colony of bats perfectly. Before dawn each day they fluttered to a small hole near the peak at one end of the roof. One at a time, they alighted and crawled quickly inside the building, then down one wall to favorite roosting places. Each bat was slightly heavier than it had been at nine o'clock the evening before. During the night it had filled its stomach with small insects of many kinds, catching every one of them in flight, individually. None of the insects had been heavy; yet the weight of so many totaled nearly an ounce. That was a lot for a bat with a body the size of a mouse.

The bats uttered chittering sounds as they entered the barn. They kept up this conversation until each had found its particular place to cling. There the bats remained all day, suspended upside-down. They slept as long as fifteen hours each day while continuing to digest the insects they had caught. Every few hours the bats awakened just enough to rid themselves of wastes. The material accumulated on the barn floor below. Fine threads of white mold grew on the wastes and gave off a musty odor.

One day a biologist arrived at the farm. He introduced himself to the retired couple who lived in the house and told them he was interested in the behavior of bats. He asked if he might look in their barn to see if any were roosting there. He would like to catch as many bats as he could, attach numbered metal markers to them, and release them all unharmed. The owners of the farm hadn't noticed any bats around. They hardly used the barn. The couple told the biologist he was welcome to inspect the barn any way he liked and to have any bats he found.

The biologist recognized the particular musty smell as soon as he opened the barn door. He looked until he found the moldy mound of wastes. He shone his flashlight on the wall above, to see if any bats were there that day. The unfamiliar light wakened dozens of the creatures. They chittered to one another, waking up those that were still asleep. But their roosting area was so high on the wall that the biologist could not reach them. He turned off his flashlight so as not to disturb the bats anymore and quietly left the barn. He shut the door tightly and asked the owners for permission to return in a day or two with a net, a ladder, and some assistants. He would be unable to catch and band so many bats by himself.

The bats had not been asleep long the next morning before the barn door opened quietly and several people entered. They brought a ladder and a large box of equipment. The people closed the barn door, shutting out the sunlight, and turned no flashlights on the bats. Yet the bats could not go back to sleep because, in the dim illumination of daylight through a small window, the people were moving about and whispering to one another. They put the ladder against one wall and climbed it. They spread out some old boards to make a crude floor on top of the tie beams and stretched an extremely fine net across the space between this floor and the barn roof. They hauled up some boxes and gloves and long-handled butterfly nets. Two of the people stayed up near the roof, while the others climbed down and moved the ladder.

Now the bats had something to excite them. The end of the ladder reached right into their roosting area, and the biologist began climbing it as quickly as he could, waving one arm and shouting. Every bat took to its wings. Just one of them remembered to escape through the small hole to the world outside. The rest flew back and forth, high up under the roof. More and more of them got caught in the fine net, and some were captured in butterfly nets. In a few minutes not a flying bat remained. All were inside the boxes, unhurt, crowded in ways they had never been before. They had no space to spread their wings. All they could do was cling to the wire screens that served as the side walls of the boxes, or climb on one another.

Some of the bats were mothers that had given birth just a few days earlier. Each one had a baby bat clinging to the short fur of her undersurface. Even with all the scrambling around inside the boxes, the babies held on tightly. None of them had their eyes open yet; they could only hear and feel all the commotion.

The people discovered that some of the bats had experienced a similar event before. These bats wore numbered aluminum clips shaped like the letter C. Each clip had been almost closed around the part of the wing that corresponds to a human's upper arm. It could not fall off nor damage the wing, and its weight was too slight to interfere in any way with flying or catching food or clinging to a barn wall all day. The biologist and his helpers knew that an even-numbered band on the left wing marked a female; an odd-numbered band on the right wing, a male. No older males were in the barn that day; they were clustered in a bachelor group in another hideaway. All of the males in the barn were young bats born the year before, still associating with their mothers and other female bats.

Outside the barn, the biologist and his helpers kept very busy. With gloved hands they removed one bat at a time from a box, read its band if it had one, and recorded the information. Released at once, the bats flew away. They did not return to the hole in the barn. Every unbanded bat was examined to learn its sex and then was given a band, even-numbered or odd, on the proper wing. If a mother had a baby, it also received a band of the same small size. By the time the baby could fly, it would find no inconvenience in the added weight. Now the mother could go free with her baby, both of them marked for later recognition.

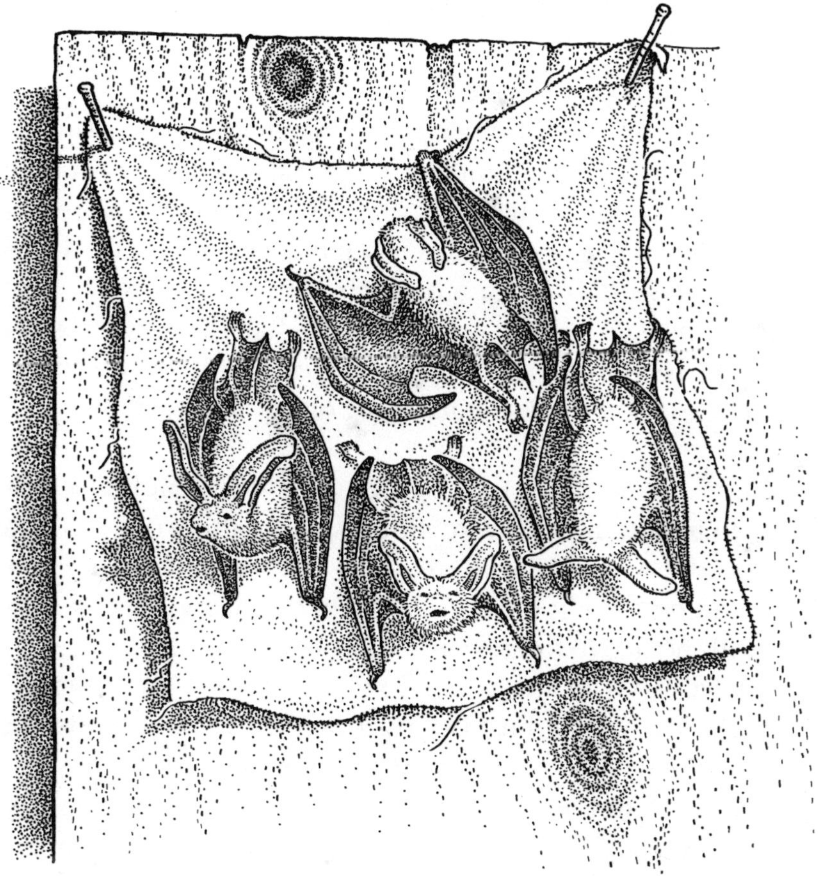

Young *Plecotus* bats

After all of the full-grown bats had been removed from the boxes, four babies remained inside. Each somehow had lost its grip on its mother's fur and had not found her again amid the confusion. The biologist banded all four, then gave them a chance to call their mothers. He leaned a piece of plywood against the outside of the barn, tacked some coarse cloth to the wood, and hung the little bats on the cloth as though they were medals. He could see their tiny mouths open but could hear no sound. Their calls were too high-pitched for human ears. The biologist stepped back, and so did all of his assistants, except one who held a butterfly net. Soon four adult bats swooped low, one at a time. Each one picked a baby bat off the cloth and started to flap away. Every mother recovered her own youngster, recognizing it by its ultrasonic cries. Without missing, the assistant caught each mother and baby in the butterfly net. The mother's tag and the baby's band were both read and recorded, so that the relationship between the two would be known. The bats flew off as soon as they were released.

The bats did not return to the barn, so the biologist began looking for them in all directions. On the third day he visited an old tree with a hollow center, where once before he had found a heap of bat droppings. Then the droppings had been dry, with no active fungus and no musty smell. This time there was a fresh covering and a few streaks of new mold. The biologist knew bats must be roosting high in the hollow heart of the old tree. Catching any of them would be far more difficult than in the barn; yet doing so would be the only way to learn if they were the same bats caught in the barn, or different ones. The biologist went back to his car and got some equipment like that used by telephone repairmen to climb a pole that has no metal rungs. With the end of a rope tied to his belt, he carefully clambered up the tree to the one place where he could see a knothole. He felt sure the bats entered and departed through that opening. He hauled up the rope that was tied to his belt. Its lower end was tied to a kit containing various supplies. With the kit in hand, he removed a few bits of wire, a hammer, some staples, and a square of mosquito netting. He built a small frame to support the netting like a bag or a small tent outside the knothole. If any of the bats came out, they might stay in the bag until he could climb again—this time at night—to see which bats they were.

Only eleven bats came out of the knothole and found themselves caged in. One turned back inside the tree, where the rest of the colony was moving about; it could still hear the bats in the net bag outside. The trapped bats continued to make sounds that told of their plight. All the other bats departed by way of a different knothole, one the biologist had not discovered. Eight of the ten bats caught in the net bag that night wore tags. The biologist marked down the numbers, put bands on the other two, and freed them all. He could hardly wait to get to the ground, where he could look up the numbers of the eight banded bats. Once there, he satisfied himself that these were the same bats he had captured in the barn.

Not a single bat returned to the hollow tree. The creatures did not go back to the barn either. Instead, after they had caught a stomachful of insects apiece, the bats assembled at the opening of an old mine shaft. It was partway up a hillside about a mile from the barn and slightly farther from the tree where they had roosted the night before. The bats came to this same place when winter made more exposed hideaways too chilly, even in California. It was farther from the valley and the reservoir where insects were easiest to catch. But when their favorite roosts were no longer suitable, it was the place for the bats to go.

It took the biologist four days to find the bats again. He inspected half a dozen hollow trees, a church belfry, and two abandoned mine openings before he reached the right one. Three of these refuges hid other bats. The biologist wrote himself a note to look into these roosts at a later time. For the moment his goal was to locate the bats from the barn and to learn whether they stayed together in some third hiding place. Already he was thinking of them as "his" bats, because nearly all of them wore wing tags he had attached. For this reason he felt especially happy when he shone his flashlight into the darkness of the third mine shaft and saw three bats of the right kind clinging to the wall. He quickly turned off the lamp, hurried to his car, and sought his assistants' help. He wanted to build a net across the mine opening before sunset.

The majority of the bats were too deep in the dark mine to be disturbed by the preparations outside. These flying mammals needed no one to tell them when night came and insects were flitting about, as though waiting to be eaten. Each bat flapped its membranous wings dexterously in total darkness while following the mine shaft in the earth. So well did the bats respond to the echoes of their own ultrasonic cries, as they bounced off the mine walls and off other bats, that collisions were unheard of. Yet that evening, when the bats emerged from the mine shaft, they struck the fine netting before discovering that it was there. Every one of the bats was caught and examined before being freed again. Only two lacked wing tags until that night. When the last flier had vanished into the darkness outside, to forget its terror and fill its stomach, the biologist turned to his record book. About a third of the baby bats were missing, although their mothers had been accounted for. So the biologist and his helpers put on headbands to hold their flashlights, took along a kit of extra bands and tools, and let themselves down into the old mine.

The searchers found the babies. Every one clung to the wall of the passageway about fifty feet in from the entrance.

Each baby had been left there by its mother while she went off for food. In so few days the babies had grown considerably. Now they weighed too much for their mothers to carry them while pursuing insects in the dark. Ordinarily, each mother would be back before dawn with food or with her rich milk. But the biologist wondered if the mothers ever would come back, now that they had been caught while emerging from the mine. It might seem too risky to return to tend the babies. The biologist and his crew went home that night and decided to revisit the mine the following day to learn the fate of the young bats.

In the mine shaft the next day there were no bats of any age or kind. Evidently the mothers had returned, picked their babies from the wall, and flown to some other place of refuge. The biologist checked earlier hideaways: no bats in the hollow tree and none in the barn. Late that afternoon he inspected the last place on his list—the belfry. He waited until the clock struck six, letting the booming sound of the big bell quiet before he climbed the ladder to the topmost level. There he saw bats—bats of the kind he sought. He managed to catch four with a butterfly net. All the rest flew away at once, although the sun had not yet set. For a few minutes the biologist was so busy recording the numbers of the wing tags in his notebook that he forgot about the bats outdoors. In daylight a hawk might see them and catch a few. He felt sorry to have put "his" bats in danger, yet he was happy to have found their fourth hideaway. What a noisy place to sleep, with that bell ringing every hour! The sound hurt his ears when he was in the belfry, so he thought it must cause pain to the bats, for their hearing is so sensitive. On the other hand, maybe the low note of the big bell just shook the bats without their feeling distress from the sound. Since his ears could not detect the bats' ultrasonic cries, perhaps their ears were deaf to low-pitched vibrations.

After the discovery of their belfry roost, the bats hid temporarily in many places. Each was suitable for an emergency but not for continued use. A few hours after darkness fell, all of the bats went hunting. Before dawn their hunger no longer drove them onward. An old memory drew the colony to a different mine shaft. Although some of the bats had never visited the place before, their delicate hearing helped them pick up the message from those that knew the way from previous years. Chittering for nearly an hour, the bats eventually found places to cling and settled for the day. Some of the mothers took a little longer to settle, for they insisted that their babies hold to the wall beside them. A mother bat can scarcely sleep with such a heavy youngster dragging on her fur.

The biologist once again inspected all the places where he had found "his" bats before. Each of these hideaways was deserted. Then he tried to imagine where the bats might have hidden themselves and made a list of unoccupied locations that seemed suitable, even though he had found no bats of any kind there previously. The mine shaft where the bats actually had gone happened to be the first location on his list. He arrived there before sunset and found them. With his butterfly net he caught just a few to assure himself that these were the right individuals, but that was enough to disturb the entire colony. The bats did not return to the mine shaft after that night's foraging. Instead, they went back to the barn.

The tagged bats showed the biologist what he wanted to learn. The colony had five different places to hide by day. When disturbed at one roost, they moved to the next. This behavior kept them together; it saved them from many kinds of danger that bats meet in a normal lifetime.

Bats may not fly far, for they are really stick-at-homes. They always return, although they know several places to hide.

North American beaver *(Castor canadensis)*

6

The Mother Beaver and Her Lodge

WHEN YOU SEE a mother beaver swimming alongside the dam that holds the beaver pond, you have a chance to sense the world as she does. The wet fur on her broad body appears almost black, not brown as it is when dry. Her flat tail is ready to rise and slap down on the water's surface if she is alarmed in any way. One splash is enough to make every beaver in the pond dive for safety.

The beaver pauses to examine any stick that projects from the mud of the dam, for occasionally a repair is needed. She may clamp down on the wood with her strong orange teeth, which are shaped like curved chisels, and clip off the stick to a better length. She listens for running water, which tells her that a leak may be enlarging in the dam, and she inspects every hole. Small leaks do not matter, but the dam must hold if she and her family are to be secure. Only after a rainstorm upstream should water flow over the top of the dam. Otherwise the water escapes through insignificant gaps in the dam at the same rate as it streams into the opposite side of the pond.

This industrious mother beaver may live to be nineteen years old. By the time she was eight she already was in her prime. By then she no longer remembered the night she first arrived at the site of her first home away from her parents. That was a few weeks before her second birthday. Never before that night had she seen this new stretch of river, and everything frightened her because it was so strange. She had come here only because her parents had driven her away. They lived nearly a mile upstream, where she had grown up, and had made her find a place to live on her own.

The night the young female arrived at her new place on the river, the water was gurgling and splashing over stones and boulders. She was still investigating the stretch of rapid water when a male beaver approached. He, too, had been thrust out from his parents' territory, but in a different river. He had been wandering for two weeks, trying to find an unoccupied region that was suitable for beavers. Discovering none and being pursued so often, he climbed over a hill and came down the other side. That was how he happened to meet the young female. She was just his age, smelled good to him, and did not bite. He helped her to explore the rapids she had found.

Every beaver inherits the knowledge of what to do when it encounters water that is flowing rapidly. Generally, the male shows that he is more daring. He crawls out on the bank where a poplar tree overhangs the water. Even though the trunk may be ten inches in diameter, he bites into it about fifteen inches from the ground. He cuts out great chips of wood and lets them fly in all directions. Suddenly the tree creaks and begins to fall. The beaver runs for the water. Often the tree topples into the river almost on top of him.

Usually some of the wood fibers still hold the tree trunk to the part that is rooted in the ground. The beaver goes back to examine the tree and uses his teeth to cut it loose. Meanwhile, the female beaver may be calmly eating the foliage and slender branches that have fallen into the water. Before long the male joins her at this feast. Gradually the two beavers cut off almost everything that projects from the main trunk. Afterward, one beaver is likely to take the smaller end of the fallen tree between its strong teeth and pull. With luck the tree will become firmly stuck crosswise on the river, perhaps extending two-thirds of the distance to the farther side. It thus provides a good beginning toward a dam.

Trimming the tree

On successive nights, the two beavers fell and trim other trees along the river bank. Most will be poplars, with some young beech and slender oak. The current can swing the bare lengths of wood or carry them downstream. Some are likely to catch on the first tree the beavers felled and form a loose barrier. Any barricade across flowing water seems to encourage the animals to work harder. Hurriedly they carry more and more branches and mud. They create such a tangle of stout sticks that the water cannot wash away the mud as rapidly as the beavers bring more. Wherever the animals find a hole, they shove in more branches and more mud until they no longer can hear a rush of water. Eventually the barricade becomes a dam. It may create a pond where none has been for more than a hundred years.

Beavers do all they can to make their dam hold firm, despite the pressure of the pond water against it. They create a private lake. Near its center or to one side, they locate a shallow place, to which they haul sticks and mud for constructing a lodge. The dwelling will be almost exactly like the lodges in which the beavers themselves were born. Water surrounds it even in the driest part of summer. Inside are two rooms. One is bigger than the other, and slightly higher and drier. The lower room is primarily a place where the beavers can shake water from their fur after swimming into the lodge through its underwater doorway. The beavers leave no place where their home can be invaded from above. Poles and sticks reinforce the mud dome, which is at least a foot thick everywhere. It is even thicker around the sides, below the level of the water.

Usually a female beaver is nearly four years old when she gives birth to her first babies in the lodge. The night before the wonderful event, she expels her mate. She insists on managing alone. She relaxes and waits; then contractions inside her body push out a helpless kit. She licks her baby clean and cuts its umbilical cord with her teeth. She does the same for the next kit when it appears. And for a third and a fourth. Again she lies on her back and relaxes. The kits nuzzle against her warm body. Eventually each kit finds a nipple and settles down for its first meal of milk. All the kits suck until their round bodies can hold no more; then it is time for all the inhabitants of the lodge to go to sleep.

The male beaver returns to the lodge late at night. He

shakes off the water from his fur and starts to crawl up to the higher, larger room. The female meets him and resists every move he makes for a few minutes. He keeps sniffing at all the strange smells inside the lodge. Gradually the mother allows him to approach the kits, smell them, and touch them with his blunt nose. He must learn that they are his kits too. He can guard them in the lodge, like a babysitter, while she goes out to fill her stomach. Protection is all the kits receive from him until, in six months or so, they are weaned. Then the mother will lead them out of the lodge for the first time. Her mate can help her introduce the kits to various plants that are good to eat.

The female beaver becomes a mother every spring. Years pass. By the time she is eight years old, she has had four litters. The last kits to be born will share the lodge with her and with the much larger young of the year before. Although all of them have the same mother, each litter may have a different father, making the youngest only half-brothers and half-sisters of those that are between one and two years old.

The father of her first litters might have ventured too far from the pond in search of a big poplar tree to fell and lost his life to a bobcat. The cat could rip him open, eat a little of his sweet meat, and leave everything else for skunks and beetles to clean up. For the rest of that summer, the mother beaver would have to risk leaving her nurslings in the lodge whenever she swam out to find food. Her latest litter then would have only the questionable care provided by the kits from the year before, who will still live in the lodge.

Before too long the mother beaver would accept the presence, and then the attention, of a younger male. He might have spent his third and fourth year wandering in search of a place to live and a mate. By then, experience would have taught him to take care of himself.

A mother beaver gets plenty of experience also, but she likes having a family around her. She expects a crowd of friendly beavers—her family and her attendant male—inside the lodge each day. Only in spring does she want solitude, just before giving birth to another litter.

The female's mate always is willing to remain in the lodge with her nurslings while she goes out at night to get a stomachful of plant foods. The older youngsters want to accompany her. They imitate her and learn beaver ways, no matter what she does. They follow and try to help each time she crawls up on the lodge before going in. She inspects her roof almost every night and the dam nearly as often. She keeps repairing them, whether they really need it or not.

Only occasionally does a beaver family seem threatened by invasion through the roof of their lodge. Then the occupants huddle together more tightly than usual while they listen to all the scratching, snapping, and snuffling sounds above them. A black bear may be trying with brute force to tear a hole through the roof, to reach the beavers inside. He can smell them, and hunger makes him desperate. Yet incredible numbers of sticks and barkless branches keep catching on his claws as he rakes at the dried mud. All of this wood has the beaver scent, absorbed while the animals ate off the bark. Their odor comes from the mud, too, for they carried it in their mouths. Finally the bear has to give up. He swims to shore and walks off through the woodland. After so much work, he is hungrier than ever.

The beavers wait inside the lodge for hours following the bear's departure before they swim silently into the pond. They return after eating a good meal. Generally the mother beaver leads her family up the side of the lodge to where the bear was digging. The beavers plaster on an extra amount of mud and sticks, making the roof thicker and stronger than it was before.

The underwater doorway is the only opening through which an animal can enter the lodge. A mink, an otter, or even a muskrat might slip in there and attack the nursling kits during the summer months if they have no protector. Usually an older beaver stays alert in the perpetual darkness where the lower, shake-dry room comes up a step to the main chamber of the lodge. In one lodge, a mother beaver decapitated a hungry mink that entered, and nipped the tail off another as it tried to escape; her mate slashed at a muskrat that came in. Otherwise, the beavers' fortress of mud and sticks kept out invaders and will continue to provide protection so long as the mother beaver keeps repairing her lodge and dam.

Each autumn, beavers spend many nights felling poplars and other trees. They cut branches and haul them to a deep place in the pond as close as possible to the lodge. There, with all the force the animals can exert individually, they shove the cut ends into the muddy bottom. A miniature forest, complete with leaves, develops from this activity. It will be the beavers' food supply for the winter months, when cold weather creates a roof of ice across the pond. The beavers swim under the ice to their food supply, loosen a branch or two, and tow the food to the lodge. The lower, shake-dry room serves as their dining room. The beavers crouch there, devouring the bark and twigs, as well as any leaves that remain attached to the branches. If a central part of a branch remains, like a hard core of wood, a beaver swims off with it and releases it in the water. Often the current carries these waste pieces to the dam, where they catch in the tangle of sticks already there.

The lodge needs no repairs all winter. From December until April, the mother beaver and her family can relax indoors under a dome of frozen mud as hard as concrete. Snow usually covers the mud and sticks, and tracks of animals often mark the snow. The size and shape of the footprints reveal whether the traveler was a grouse, a snowshoe hare, or a squirrel. A fox creates different patterns according to whether it is walking quietly or leaping ahead to catch some creature. An ermine, which is a weasel in its winter coat, always leaps, always seems to be in a hurry. These comings and goings across the pond, to the top of the beaver lodge and down again, never disturb the residents inside. They have their food supply and one another. Each one has its heavy coat of fur, which holds a blanket of warm air next to its body, as insulation against the cold. The river provides the beavers with a great many bubbles of fresh air and carries away their wastes. The animals feel no need for anything more.

From April until December, the mother beaver keeps her lodge in good repair. Her family helps her maintain the dam. This cooperative activity keeps the beavers secure against animal-eaters and the rushing river in spring and fall. Each year the two-year-olds in the family have to be driven off, to set up beaver ponds and lodges elsewhere if they can. These emigrants keep the world supplied with beavers in as many places as are suitable for them. Each beaver pond sustains an almost constant population; consequently, the single family encroaches only very slowly on its food supply.

The thriving beaver family cannot continue indefinitely to send out young beavers every year. Even if the mother stays healthy until she is nineteen years old, other factors come into play that reduce the size of the group. The river waters carry a load of silt each time the river floods. The silt fills the pond a little more each time, and shallow areas expand. Plants grow in these areas—plants that beavers like to eat each summer. But land animals that do not dive for food can walk over the shallows, nearer and nearer to the lodge. Deep water becomes more restricted and located farther from shore. And it is only deep water that allows beavers to submerge for safety and store their winter food supply. Moreover, when all the edible trees near a pond have been cut, they take years to regenerate. So every autumn the beavers must go farther onto land to get branches to store for the winter. While the beavers are at work, bobcats and other beaver-eaters are likely to ambush them. Thus, the beaver family shrinks in size.

With fewer helpers, the mother beaver cannot store as much food in the mud at the bottom of the pond, and she and her family may be near starvation before spring. They may be unable to prevent meltwater and floating cakes of ice from destroying their dam. If a small hole begins to let flood waters through, the whole structure may break apart. If the pond drains, the lodge is exposed completely.

Sometimes a mother beaver seems to recognize such danger. She leads her family upstream or down, away from her old lodge and the shallow pond, away from all familiar territory. She knows she must find a better food supply and a place to build a new dam and lodge or all the beavers will perish.

Wherever an old dam collapses and a beaver pond vanishes, new plants grow on the exposed silt. Shrubby bushes get taller every summer, and trees near the old pond shores rise again and cast more shade. The river cuts a narrow channel through the silt. After a while, which may be anywhere from ten years to a hundred, the river is gurgling over the rocks again. It makes the sound that appeals to any young female beaver, before she is a mother or even has a mate.

Soon an industrious beaver comes along, a family outcast looking for a home site. A mate arrives. A new dam is built, changing the turbulent river into a tranquil pond. The beavers begin to build a lodge surrounded on all sides by water, in the habitat that suits their special needs. Everything seems the same as it was a century before. Only the beavers themselves are different, although they act in the traditional way.

Snowshoe (or Varying) hare *(Lepus americanus)*

7

The Hare That Never Gets Lost

LONG BROWN LASHES concealed the dark eyes of the hare. It watched attentively as the red fox came along the forest path. Hunger kept the fox alert to every scent, every sound, and every movement within sight. But the hare did not move a whisker. Even its breathing movements ceased while the fox walked by, just a few feet away. To the fox, the hare appeared to be merely an inedible rock, a solid mound about eighteen inches long and eight inches high. It was brown flecked with black, odorless, silent, and unworthy of investigation. The fox continued on its way and disappeared. The hare twitched its nose and breathed again but kept its black-tipped ears laid smoothly atop its back. Its body concealed its feet, which were pressed firmly against the ground.

Day or night, a hare must be ready to leap into action. Its long legs can propel it four yards in a single jump. For a dozen jumps in quick succession, thirty miles per hour is normal speed. The next leap may be an extra-high one; this enables the hare to look back and see if it is being pursued. By then the hare will be turning in a new direction to escape danger. Generally it curves around toward its starting point. If really frightened, the hare twists its body in mid-leap and goes off at a sharp angle. Chasing a hare is exhausting work for any beast of prey.

Any time you discover a hare in an open woodland, you can repeat the test that a biologist tried in Wisconsin. The hare he found rushed away but stopped before disappearing from sight. The biologist looked at his watch, got out his compass, and walked slowly after the hare. The animal took off again, to the right of its original route. Compass in hand, the biologist moved even farther to the right and soon induced the hare to go to the left. Gradually, so as not to alarm the animal too much, he drove it in one general direction. For a while the plan seemed to work; then the hare turned and rushed past, back into the area from which the two had come. The biologist circled, attempting to get the hare headed—even zigzagging—into the chosen course. It was no use. For an hour and a quarter he followed the hare. He kept it moving, but never beyond a small area about eighty yards in diameter. That area was home to the hare, and it refused to go one jump outside its familiar territory.

Any hare that lives to be four or five years old knows its home area thoroughly. Every rock, tree, and bush is a reliable feature of the territory. The animal travels some routes so often that they become visible trails, marked by bare ground or by weeds that tolerate being stepped on repeatedly. Across a peat bog or some other mat of moss, the trail may be three inches deep. The hare hunches itself along at slow speed while browsing on buds and twigs. It leaps from danger over the same routes. Without thinking, it relies upon memory and can hurry even in total darkness—three jumps, then right, one leap, now right again, six jumps, and to the left—without mistake. The animal does not need to see the obstacles it avoids by the routine it follows so frequently. It knows exactly where it should vault over a fallen log and where it can get off any trail to a suitable crouching place. Any pursuer moves along more cautiously and is likely to pass right by.

Each hare has its favorite place to spend sunny hours. It may have scraped a slight depression in the ground, as though aware that this allows its body to sink lower, hiding even its shadow. Generally the place offers a fine view in all directions and some protection from the sun under an overhanging leafy branch. Unless disturbed, the hare will stay here until the sun sets. At that point the animal stretches as if its legs were stiff. It may sit back on its haunches. Sometimes it rises on hind legs alone and paws at the air with its front legs, perhaps to help balance. Then the hare goes down on all fours and hops to the nearest path. It moves slowly, a few steps at a time. As the hare brings forward the large hind feet that support most of its body weight, each one is set down at the side of its smaller front feet. Next the animal lurches forward and uses its front feet to share the burden. Again it may hop to bring its hind feet into the same general location as the front ones.

When a hare is in a hurry, its footwork is quite different. Without pausing, it swings its hind legs well past the front pair. It flops them down on each side many inches ahead of its front paws. Like a warm projectile, the hare's four-pound body continues on. Its front feet stretch forward to touch the ground, keeping the head up, while the hind legs provide another powerful thrust. On soft mud or snow, the track becomes

spaced thirty-six to forty-eight inches apart, instead of the customary

of the hare while feeding.

A mother hare barely stops traveling to give birth. When the moment comes, she raises herself by extending her hind legs almost vertically, and the baby slides out in its protective sac. She turns slightly and cuts the sac open with her teeth. The youngster crawls out, with both eyes open. The mother eats the sac and licks her baby until it is clean and dry. Then she hops ahead, and the newborn hare runs after her. She stops to let it nurse a while. Within minutes a second baby is ready to be born, and she repeats her procedure. The hare's entire litter, which may number from three to eight, makes its appearance within half an hour, and the mother will have moved twenty to thirty yards farther along her path from the place where her first baby was born. Every youngster follows close behind her, waiting for another chance to get some milk.

Unlike rabbits, hares make no burrow, no nest of any kind. By the time a young hare is a week old, it can raise its ears and hop. Before the end of the next week, it is sampling some of the same plants its mother eats. By contrast, a young rabbit of equal age still will be in a fur-lined nest, not yet ready to open its eyes for the first time. Even when fully grown, a rabbit's legs are never as long as those that carry a hare forward at such great speeds. And only the hare grows a breadth of bristly hairs upon its large hind feet, which leave a mark in snow suggesting the imprint of snowshoes. This is why the hare so often is called the "snowshoe hare"—or the "snowshoe rabbit" by people who are unaware of the difference.

A mother hare takes few chances with her babies until they are old enough to be independent. She leaves them individually in locations she regards as safe while she hops off to browse or rushes at top speed to elude some pursuer. When her stomach is full and the way seems clear, she returns to each youngster and lets it nurse in its hideaway. A baby hare weighs about a quarter of a pound at birth and gains about half an ounce every day. Within a month it will wander off, completely on its own. It will show no sign of recognizing its own mother if it meets her elsewhere on her territory. After all, she is not the only hare in the ten-acre area. Other adult hares visit because their home lands overlap, and their youngsters, too, will be exploring, to get used to the location of obstacles and food. Snowshoe hares find what they need in many places between the Pacific and Atlantic oceans and as far north as there are trees. A northern tract of spruce and fir with boggy areas suits them just as well as a thicket of swampy land somewhat farther south.

Male hares cover more ground than females and try to find a female as soon as she gives birth. The female is willing to mate again once she has cleaned and fed the last of her litter. This explains why she can produce three litters of young each summer at less than six-week intervals. Only the first courtship of the year seems more of a game, with the female leading the male on a merry chase through the forest, dodging him and spurning his advances. From March into April, she refuses to let him catch her, or she kicks and bites at him if he comes too close. Suddenly she is ready and accepts any male approaching her until her mood changes.

Hare on a snow elevator

In autumn the hares seem more subdued. They are too busy eating, adding to their reserves of fat, to show much interest in other activities. Winter is coming. A fat hare has nourishment to live on while storms rage. Snow offers no real hazard. The snowshoe hare can walk on top of all but the most powdery snow. The white covering over the earth serves, in fact, as an elevator. It raises the hare to levels of branches and bark that it could not reach while standing, even tiptoe, on the ground.

During the autumn months, the hares grow a new coat of fur and shed the old. The new covering is thicker and snow-white. It keeps out the cold and transforms the brown camouflage of summer into an equally effective masking device when snow is on the forest floor. Not until spring will brown hairs reappear, as the hare gets ready for warmer weather. Then the white hairs are shed, and the hare becomes the color of dead leaves. Its ability to vary in color, from brown to white and back to brown again, explains the name "varying hare" for the snowshoe hare. Curiously, in parts of Washington State and British Columbia where snow rarely covers the ground for more than a few days at a time in winter, the hares remain brown year round. For their locality, it does not pay to vary. A white hare would be conspicuous most of the winter, but a brown one can wait motionless for the temporary covering of snow to melt.

A young hare needs months of experience to learn whether it must run for its life or can remain safe merely by staying completely still. It takes months to train itself to leap ahead of a pursuer without crashing into some obstacle. During these months, many a young hare becomes a meal for a bobcat, lynx, fox, coyote, or, in some places, a wolf. At night the hare may be pounced on silently by a great horned owl or a great gray owl. Consequently, only about thirty out of every one hundred young hares survive to their first birthday. The thirty survivors mature but still face many dangers. Fifteen of them disappear before they are two years old, and eight will be lucky to reach age three. Rarely do more than four live to their fourth birthday. Two may be left at five years after birth.

Hares have no way to protect their young and themselves from diseases or to find real safety from birds and beasts of prey. They gain what security is possible by staying close to home, in familiar territory. They never get lost. Rarely does one die from starvation, for there is food all around it; yet each move to get something to eat exposes the hare to danger. Escaping is its way of life.

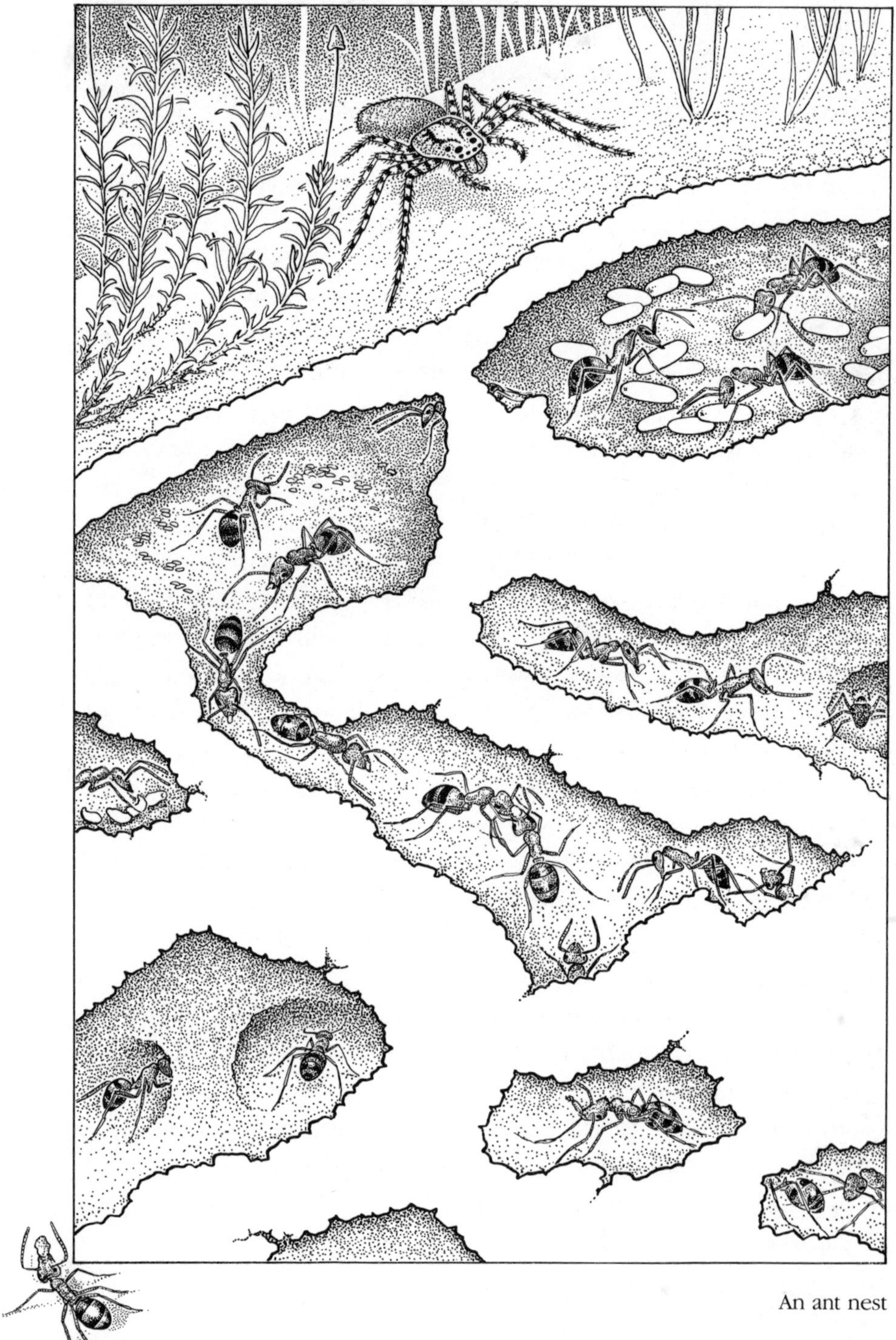

An ant nest

8

Homebody Ants and Gypsy Ants

THE BIG ANT nest is about four feet across. It rises almost fifteen inches high at one side of the pasture, next to the woodland. Under the dome live half a million ants, each nearly half an inch in length. If all of the ants stood in a single line, the line would be more than forty miles long. But not a single ant has been as far as forty feet from home. That nest is where the queen ant lives.

The queen is the mother ant. She is sixteen years old now but appears to be in perfect health. In complete darkness underground, she is tended by more than a dozen of her daughters. They crowd around her in the royal chamber. Some bring her droplets of food, a magic blend of all the kinds of nourishment that field ants carry to the nest. Other daughters wipe their soft mouthparts over the body of the queen and keep her spotlessly clean. Still others catch each pale gray egg she lays and carry it to another chamber in the nest where the egg will hatch.

On clear days, except in winter, the sun warms the surface of the ant mound. The warmth extends downward an inch or so and spreads quickly because the ants have dug dozens of small chambers, each with an earthen roof barely an eighth of an inch thick. To these chambers the ants bring the eggs that the queen has laid. They hatch sooner with the extra heat. The ants carry the grublike hatchlings to safer depths for the night and then return them to the warm nurseries the next sunny day. This helps the young digest their food faster and grow more quickly.

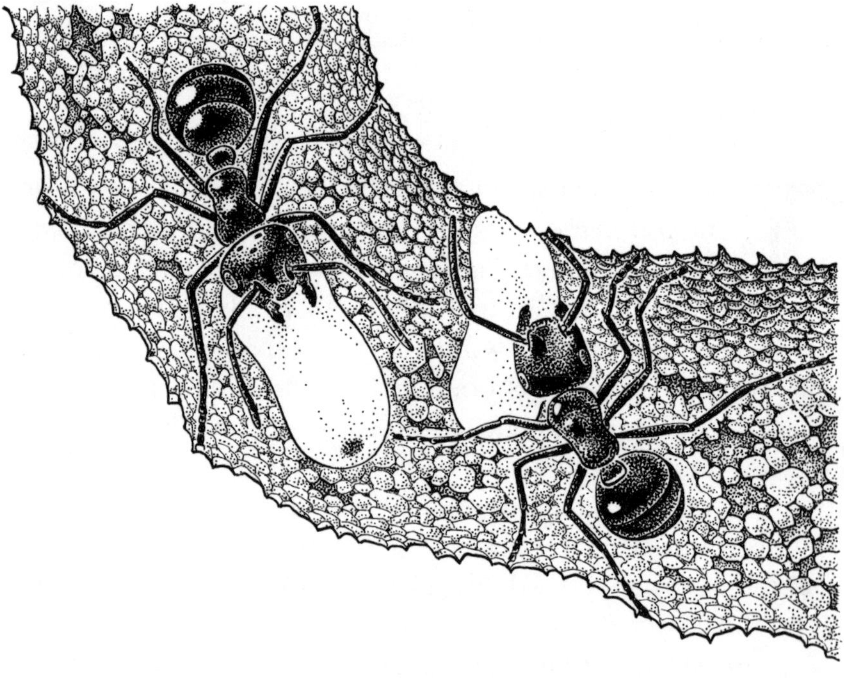

Northeastern mound-building ant *(Formica exsectoides)*

Eventually each immature ant spins a cocoon around itself. Inside the cocoon it has the privacy in which to change into an adult ant. This change occurs more rapidly if the queen's daughters carry the cocoons into the warm chambers as often as possible. Each cocoon resembles a soft grain of Puffed Rice. Its concealed occupant soon cuts its way out and becomes another working ant in the nest.

There is plenty of work to be done underground. Eggs and young must be carried up and down the nest. Food must be received from worker ants that have been foraging outdoors. Particles of earth must be removed to make more storage chambers. If a tunnel collapses, it must be cleared. If an ant dies inside the nest, the body must be dragged outside the doorway and disposed of. And most important, the queen must be cleaned and well fed.

Worker ants that venture into the open risk their lives while hunting for food. Their eyes are not nearly as good as those of the wolf spider that watches for an ant to come along. The spider pounces on an ant and injects its body with venom. The poison stops the ant's resistance and rapidly dissolves its inner organs. Once these organs turn to liquid, the spider squeezes the body of the dead ant and drinks its contents, then discards the empty covering.

Other ants explore a patch of sandy ground, only to stumble into a trap that an ant lion has constructed. The trap is a shallow pit in the sand, a cone-shaped depression perhaps two inches across and an inch deep. The ant reaches the edge of the trap, which crumbles. The unlucky insect skids downward in a sandslide. The miniature avalanche may go all the way to the bottom of the pit, or it may stop, giving the ant a chance to crawl out. But all of the slope is treacherous, and the ant skids down again. This time it ends in the jaws of the ant lion that lurks at the bottom of the cone. The ant lion seizes the ant and drags it out of sight while injecting digestive juices into its body. Soon only the empty body of the ant remains. With a flick of its head, the ant lion tosses the dead ant out of the pit and waits for another insect to arrive.

Ants in temperate lands hurry about in search of food almost exclusively after the sun rises and before it sets. The sun becomes a sort of guiding star which helps the ant find its way home again by a different route. To use the sun in this way requires only mediocre eyesight; a special memory keeps track of every turn and detour. Each ant also provides itself with a second means of getting home in case clouds hide the sun. As it travels, the insect keeps touching the tip of its body against the ground, leaving a slight streak of odorous material each time. If all else fails, the ant can find these streaks again. The insect's sensitive feelers tap against the ground and pick up the scent. The ant retraces its steps, even though the route is long and irregular, and eventually arrives home.

When an ant goes home with food, usually by the most direct path it can find, it leaves a slightly different trail of odorous materials. These chemical secretions are daubed on the ground at frequent intervals and form a series of trail markers which other ants from the nest can follow to the food source. All the ants may arrive in short order at a picnic table that a single ant has discovered.

Such a trail of ants is just what a flicker likes to find. The large woodpecker settles itself quietly on the ground at one side of the trail and eats each ant as it comes along. Still better for the bird, and worse for the ants, is for the flicker to discover the dome of the nest. The bird uses its beak to open the chambers where the ants are warming their eggs, young, and cocoons. Before the workers can carry their treasures to deeper safety, the flicker gulps down dozens of them.

Any ant risks its life when it goes outside the nest to search for food. Only a few of the many thousands in a colony take this chance at any one time. Most of the ants stay below ground, where the queen, the masses of stored food, and many of the young seem safe. Even those ants that venture outside on sunny days seldom need to hunt for hours to find trophies to carry home. The list of useful, acceptable items is almost endless. A grass seed can be grasped in the ant's jaws and doesn't interfere with running along the trail. A drop of honeydew may be sucked from an aphid that is feeding on a plant. The liquid is rich in sugars and disappears inside the ant as neatly as gasoline into the storage tank of an automobile. A little of the sugar nourishes the ant, providing it with useful energy, while the remainder will be shared with other members of the colony inside the nest. Although a dead fly, caterpillar, or some other irregular piece of meat is less convenient to transport, several ants can cooperate in hauling such a bulky prize. Or they may use their jaws to cut the meat into pieces, each small enough for an individual ant to manage.

About six hundred different kinds of ants are stick-at-homes in the United States. Each kind chooses a slightly different way of life. Some prefer open grasslands with few trees; others prefer forests with little grass; and a few thrive in deserts. Yet the ants of America north of Mexico and of temperate parts of Eurasia show far more similarity in their behavior than the nearly three thousand kinds that live in the tropics.

Most famous of the tropical ants are those that follow a gypsy lifestyle, with no permanent home of any kind. Those living in Central and South America are known as army ants, while the African ants with similar habits are called driver ants. These ants differ in ways that only a scientist is likely to appreciate. All of them move about, taking their queen along.

The queen makes this possible by staying slim enough to travel with her army for about seventeen consecutive nights. After that, the entire colony takes a three-week holiday. The first week the queen rests and grows amazingly plump. During the next week she lays eggs at a frantic pace—as many as a hundred thousand of them. She rests again during the week needed for the eggs to hatch and gets back into shape for further travel. Then off go the ants again, to a new campsite every night, a new hunting operation every day.

The queen of the army ants is slightly more than three-quarters of an inch in length. She is the largest member of the colony. The smallest of her daughters, which are scarcely over an eighth of an inch long, tend her constantly. Larger daughters, which are nearly twice the size of their smaller sisters and far more numerous, do most of the work. Still larger daughters, each about half an inch long, guard the outer boundaries of the colony. These guards possess enormous heads and hook-shaped jaws so huge that they must rely on smaller workers to feed them. At the slightest commotion, these jaws open wide. The guard bites down hard on any intruder, while stabbing repeatedly with a venomous stinger at the rear of its body.

Before dark each day during the hunting season, all of the ants prepare to move onward. Nearly a hundred thousand of the middle-sized workers find a grublike youngster to transport. Each one is picked up gently in tonglike jaws. The ant army then heads to its next destination, carrying the young along, helping the queen with her retinue to keep up.

By midnight the army finds a stopping place. The ants cluster there until dawn, in a temporary encampment that a military person would call a bivouac. The bivouac is the day's resting place for the queen, her young, her tiny attendant daughters, perhaps a thousand of the middle-sized workers, and a good many guards. The rest of the colony marches off in various directions. Big-headed guards, at intervals along both sides of the columns, appear to keep each stream of marchers in a tidy line. Each moving column has no leader to show it the way to go. Like a miniature flood four inches wide, it pours along. The ants at the leading edge seem unwilling to be pushed into unfamiliar territory; they turn and try to get back among the oncoming throng. But advancing ants shove the timid ones aside. The column discovers beetles, grasshoppers, lizards, and nestling birds that have not been able to escape. Promptly the ants overwhelm each prize. They cut it into pieces which they can carry back alongside the line of march.

The trophies move steadily toward the place where the queen, her brood, and her retinue are waiting. But much of the nourishment is devoured by advancing workers along the way. If the hunting is good, the queen and her growing young get plenty of food from the many columns. The tiny workers take their share as they prepare the food to feed the queen, the immature ants, and the big-headed guards.

Any insect that can fly may take to its wings as soon as a column of ants comes close; yet it may escape one danger only to meet another. Tropical birds of several kinds accompany the raiding ants all day. The birds chirp noisily as they dart after and eat the insects that fly up from the ground or away from some stem that the ants are investigating. Wherever the ants explore, they leave no meat behind. The workers secrete a substance that alerts other columns not to waste energy on overrunning the same place twice. But many areas along the army's route remain untouched. In them the creatures upon which the ants depend for food have a chance to rebuild their populations, even though thousands are eaten each day.

The hungry young of the army ants are the ones that keep the workers busy. The young exude a flavorsome secretion that excites the workers. The adults obtain this exciting substance while transporting the grubs each night. A worker that finds no youngster to carry goes repeatedly to lick one that some other ant is holding. When the young attain full size and spin cocoons around themselves, the workers no longer can obtain the exciting substance during the nightly travel hours. Then the workers seem to go on strike. They laze about all day, neither marching off to hunt for food nor readying themselves for a move to a new site after dark. In the first week the queen grows obese. The next week she lays her eggs. One more week is needed for the eggs to develop and for the young in the cocoons to reach adult form. Then everything happens at once: the new adults emerge from the cocoons; the eggs hatch, releasing thousands of young; and the workers once again taste the substance that urges the army into action. The time of rest and relaxation is over, and the ants begin marching off again, in different patterns by night and by day. Each night their queen is ready to go along, like a gypsy queen with daughters forming her caravan.

Among army ants the queen may live as long as five years. During the last half of her reign she produces some slightly different young. A few mature as virgin queens; like their mother, they lack wings entirely. A hundred or more become winged males and survive only a week or two. The lucky ones fly by night to other colonies that have a virgin queen waiting for a mate. A mated queen can induce some workers to help her found a new colony. Soon she replaces them with working daughters of her own. Her success is limited mostly by the hunting habits of the workers. They expend so much energy in finding and overpowering their prey that only the surplus food can go to the queen and her young. On a poor day the hunting may yield very little, while on a good one it can yield the nourishment from half a million insects, a snake or two, and extra tidbits. All of this contributes to the vigor of the entire army-ant colony.

Most other kinds of ants seem satisfied with vegetable food. They find what they need with far less effort and store their trophies in cavernous chambers, like cellars, underground. Some even raise fungi, like miniature mushrooms, on compost heaps made of leaf pieces, which their workers have cut and brought to the nest. Usually the colonies are numerous and their exploits much less dramatic than those of army ants. Their queens produce a new brood of winged females and winged males at least once each year. They are stick-at-homes instead of gadabouts.

The social habits of ants are extremely ancient, having approached perfection at least sixty million years ago. These ways of life seem to change surprisingly little as the centuries go by.

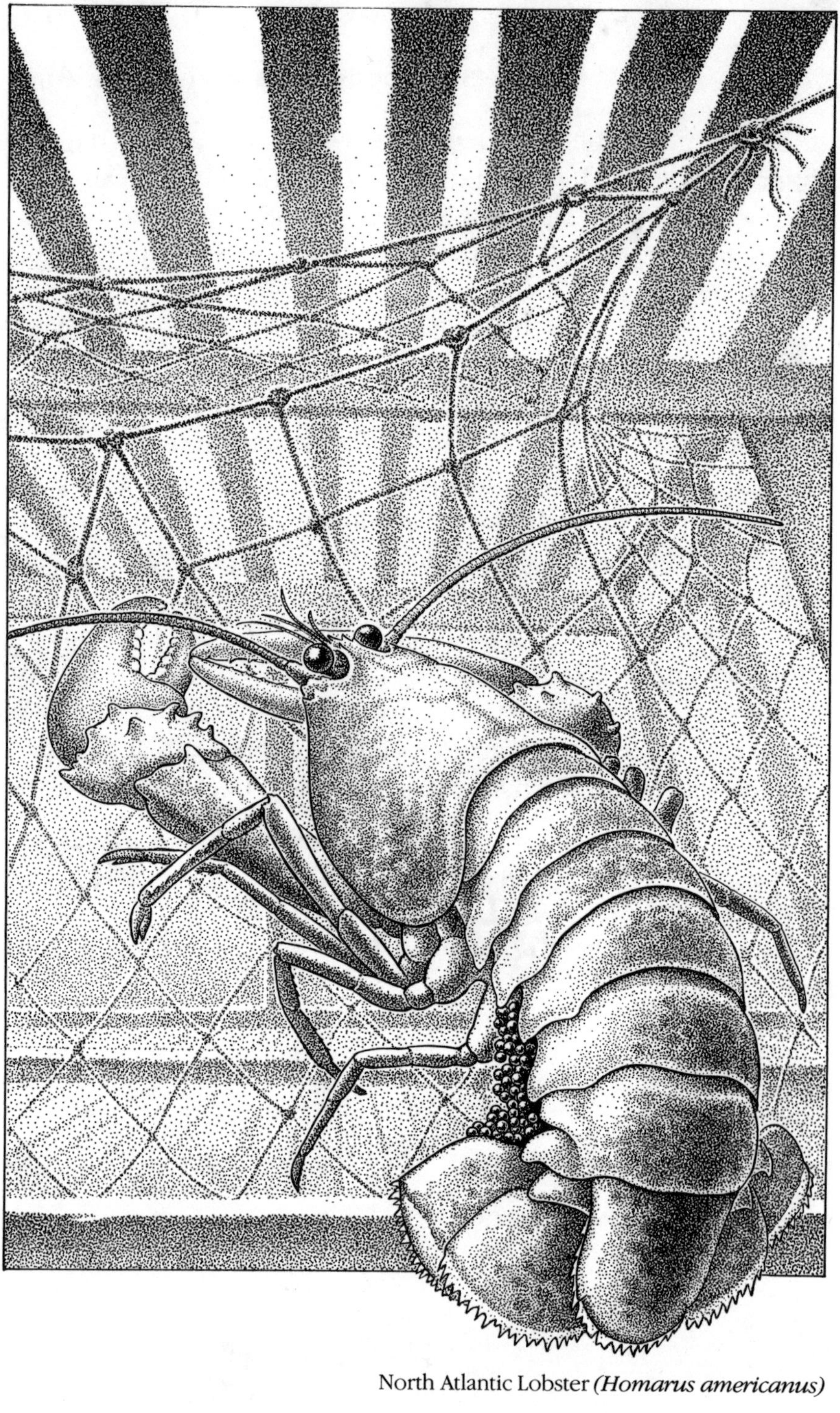

North Atlantic Lobster *(Homarus americanus)*

9

The Lobster's Two Lives

BY THE TIME a mother lobster is sixteen years and three months old, she is likely to weigh nearly two-and-a-half pounds. She may measure twelve-and-a-half inches from the end of her tail to the tip of the sharp spine that projects forward between her eyes. She is almost sure to have about 93,000 eggs glued to the four pairs of swimmerets beneath her tail. The most she can do is wave her swimmerets back and forth a little, giving the eggs the benefit of a change of sea water. Each egg is almost an eighth of an inch in diameter; yet the large number of eggs does not weigh the mother down because they have their own buoyancy in water, just as she does. In the air, the eggs would weigh about a fifth as much as the lobster's body.

A lobster this big probably has been out of water at least twice, although never by choice. The experience comes each time the lobster's scavenging on the sea floor leads her into a lobster trap. She barely hesitates as she clambers up the funnel of cord netting at the end of the trap. Passing through a ring, she reaches the central chamber and the rotting head of a codfish that is impaled upon a projection from the cage bottom. She enjoys a good meal and then cannot make her way out of the trap. Daylight finds her crouching inside, far from her customary hideaway under a big stone on the sea floor. By mid-morning the trap gets hauled up and over the side of a fishing boat. A man, pleased to have caught a lobster of legal size, opens the trap door and picks her out. Then he notices the eggs beneath her tail. With some harsh words, he heaves her overboard. By then his boat has moved on. After the lobster sinks to the sea floor, she has to walk a long way to get back to her territory again.

The mother lobster does not know how lucky she is to have a load of eggs below her body when she is caught. The eggs save her life, just as she is trying to save the lives of the 93,000 unhatched lobsters for which she is the sole guardian. She curls her tail fin around them whenever anything disturbs her. But when June comes, the eggs will hatch, and the shells will drop off her swimmerets a week later. Then the lobster will be free. By August she will be ready to molt again.

A mature lobster grows a new shell—all soft and crinkled up beneath the old one—outside the delicate living cells of its skin. When the new shell is ready, the creature bends its whole tail under it, between its legs, then stands up as high as possible inside a familiar shelter. The old shell splits crosswise at its weakest place—just behind the firm forward part of the body which lobstermen call the "barrel." The lobster struggles backward through the gaping slit and sheds every bit of its old shell. Even the thin, horny lining of the digestive tract pulls free. Ever so gently, the creature emerges, leaving the empty shell as though it were a shadow. Then the lobster rests, keeping its soft body distended with all the water it can swallow. The new shell, stretched until every crinkle is smooth, first hardens on the outside. The lobster is nearly two inches longer than before, broader in the body, and thicker too. It waits concealed for several days until the shell hardens completely. Then it ventures out to seek a meal.

Searching for food by day is dangerous, so the lobster waits until night to forage. It feels its way in the dark by wildly waving its long antennae. With its short antennae it samples the water for an inviting flavor. Sometimes the lobster finds a sea worm or a shellfish it can crush. Several pairs of mouthparts move like fingers to shove the food between the two hard jaws, one on each side. Whatever the animal swallows goes first to a crop for storage, then through a grinding mill that serves the lobster much as a gizzard does a bird. Only the softest and most minute particles move on from there through the very slender intestine. Everything else must be cast out again through the mouth.

The lobster is a stealthy hunter, alert to seize larger prey if it can. The animal holds its front pair of legs far forward, their claws open, ready to grasp any fish. The next four pairs of legs carry the lobster along lightly and enable it to swivel quickly if one claw catches something and the other claw must be used to hold the prey. But if the creature that the lobster catches proves to be large and dangerous, the lobster must be prompt to let go and disappear. It holds its tail extended far behind, ready to curl under if escape backward should prove necessary. By using its strong tail muscles, a lobster can dart astern twenty-five feet in a single second, which for that brief time is a speed of seventeen miles per hour.

Lobstermen say that herring is the best possible bait for lobster. Ordinarily, herring are far beyond a lobster's reach, swimming in schools at the surface while the lobster gropes along the bottom. Codfish feed along the bottom also, but most are too big for a lobster to attack. A young flounder is too flat, and a young shark is too fast. No wonder a hungry lobster will take a chance on entering a lobster trap to reach the bait inside.

The search for food that can be found in darkness takes each lobster many miles every year. As the water near shore gets warmer in late spring, the animals move into coastal shallows that are seldom deeper than sixty feet. When autumn storms warn of winter's approach, the lobsters move away, most as far as five or six miles from shore. There the water is 120 to 300 feet deep. Lobstermen have learned to set their traps in lines parallel to the shore and at different distances from land according to the season.

The sad fact is that storms often break the six-ply rope that links one lobster trap to the next and to a float at each end of the line. The lobsterman has no way to recover a lost trap or to do anything about the lobsters that get caught inside. Even after the bait dissolves completely, lobsters still creep into the one small opening. A succession of lobsters can die inside, each becoming bait to lure another to its doom. Such a trap goes on fishing until its slats rot away or the netting on its ends disintegrates. No one knows how many lobsters perish in this useless way.

With better luck, a lobster can live for many years. The males molt each autumn, revealing how well or poorly they have found food by how much they increase in length. An extra inch is below average, with anything more than two inches a measure of real success. But then a male has mostly his own welfare to maintain. He may find, court, and win several females at molting season, but he contributes very little energy to the next generation.

A female lobster gains lasting weight and larger size more slowly. She generally meets a mate soon after she has shed her shell. While she is unable to defend herself, he holds her down. He supplies her with enough sperm cells for a million eggs, then lets her go. It is up to her to keep the sperm alive and to get her eggs ready to lay. Almost all the nourishment she obtains from food goes into her eggs as yolk. By August the eggs occupy most of the space available inside her body. Finally she lets them out, one at a time, and ensures that each is fertilized before she cements it to a swimmeret below her tail. She then can begin to use her meals exclusively for her own nourishment.

Once she bears her eggload outside her body, the female lobster gains weight. The embryos inside the eggs are using the energy she provided in the yolk. Each one has enough to last for ten-and-a-half months of growth. During these months the mother may not be as quick on her feet as she would be without the eggs. At this time, though, she can ready herself for her next molt. The female lobster's reproduction timetable limits her to one molt every two years; thus her size does not increase with her age as rapidly as a male's does.

With each molt, a lobster gets room to grow some more before its hardened covering becomes uncomfortably tight. This process offers an extra opportunity early in the creature's life, for the new shell can differ from the old one in more than

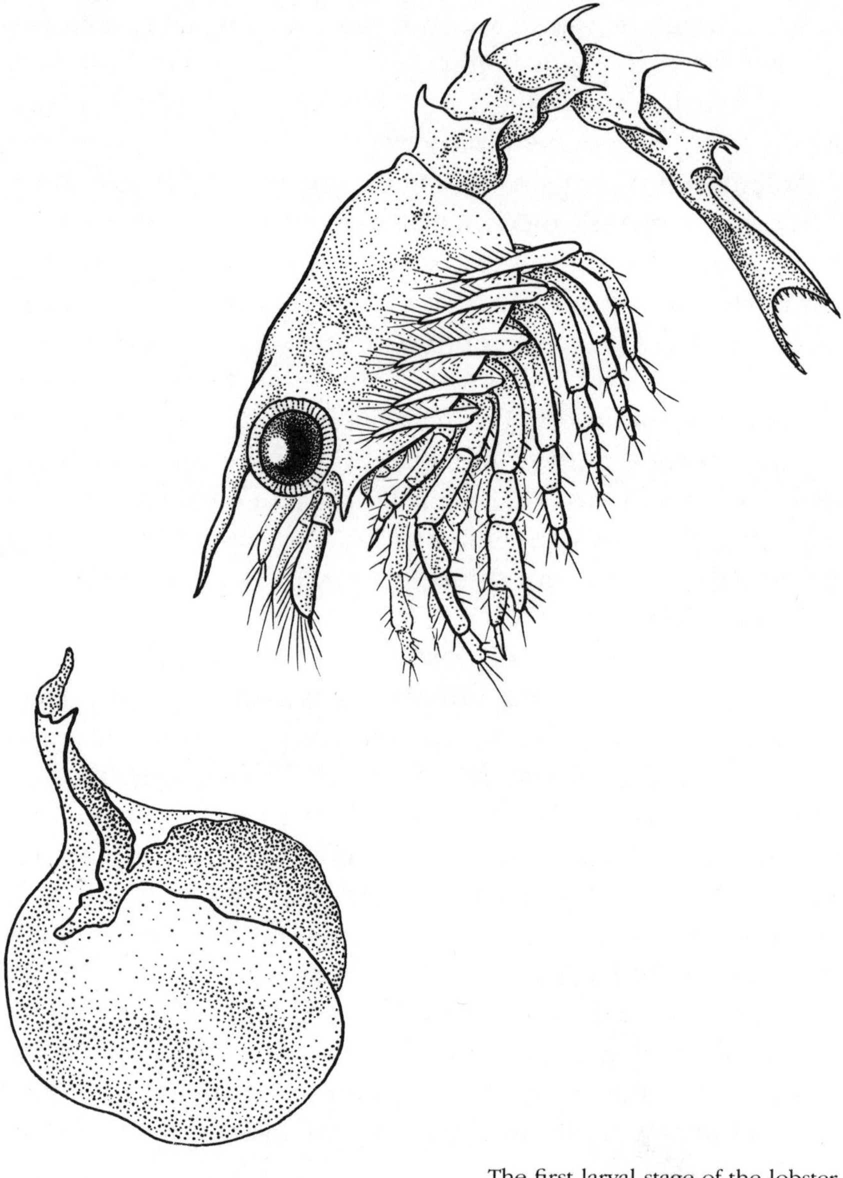

The first larval stage of the lobster

just size. Molting also permits a lobster to change its shape to match its way of life. And each lobster regularly leads two distinct lives in succession after it hatches from an egg.

The baby lobster actually is stuck inside the egg. It has to molt its shell and the eggshell to escape. The hatchling unfolds and swells its body as soon as it is free. It is about a third of an inch long and is bent at an angle where its short tail connects to the forward part of its body. The animal is lighter than the water and floats upward from where its mother lives to the surface of the sea. In less than half an hour the hatchling is among other kinds of life drifting where the sun is strong each clear summer day. Its first world is a busy place.

The little lobster is almost as transparent as colorless glass. Its beating heart and blood vessels can be seen, and so can a pair of yellowish-brown glands which are connected to its stomach and look like clusters of minute grapes. A flat, black eye marks each side of the head region and appears enormous in proportion to the size of the body. Between the eyes, a conspicuous transparent spine projects forward like a sharp spear. At the opposite end of the body is a flat, triangular plate, the last segment of the tail; the hatchling uses it to back up, much as it will its extra tailparts when it gains them after its third molt. For the present, the little lobster swims by means of five small pairs of legs and one long pair of mouthparts, as though it had six pairs of oars.

The hatchling pursues small, floating particles of all kinds. It catches and eats minute drifting plants and animals. Gradually it learns to reject objects that are useless as food. Yet its movements seem awkward, for it still has no sense organs to tell it which side is up and which is down.

The richness of the food supply in surface waters determines whether the hatchling will be ready to molt to a larger size in one day, or as many as five. When it does molt, it will increase in length by less than a fifth and display paired appendages — still hairless — beneath its tail for the first time. Its big eyes will be on stalks, more rounded, but not yet movable. The pincers on its first pair of legs, which are destined to become the prominent claws of the adult lobster, are small but useful.

The young lobster continues to drift among the living things near the ocean's surface, capturing and eating whatever creatures it can overcome. It shows no fear and swims down only a short distance when the sun is particularly bright. It retains these habits after another molt, through its third stage as a larval lobster. Less than two weeks old, it still swims mostly with its body curved and with its head and tail hanging downward. Its tail fan is complete now, and the swimmerets beneath its body have short hairs, making them effective in wafting water toward the lobster's tiny gills. Its claws are bigger, but they still hang down unless needed to help hold some struggling prey.

The next molt is more critical. After this molt, the young animal is properly a lobsterling. For the first time the creature is more than half an inch long, and its sense organs permit it to balance itself in relation to the pull of gravity. The lobsterling stretches itself horizontally in the water. It holds its bigger claws in advance of its body, for they have undergone a half twist and finally are able to grasp things from two sides at once. Its compound eyes are now on movable stalks. The animal is heavier than water and gradually sinks to the bottom. It shows signs of fear, which may save its life now that it is entering a different world. The lobsterling hunts for hideaways and backs into those that are the right size.

While daylight illuminates the sea floor, the lobsterling conceals itself in its temporary home. Just its feelers and claws are exposed, to detect and defend itself from anything that approaches quickly. Its hidden body already is becoming dark with pigment.

In spring, all of the lobsterlings begin traveling toward shore. They may need three months to get close to the edge of the water at lowest tide. By then they will have molted ten or eleven times and attained a length of one-and-a-quarter inches. They feed on worms, small snails, and anything else they can find that tastes like meat. But winter cold, and particularly the formation of ice along the coast, will drive them toward deeper water. They will follow this pattern of migration for the rest of their lives as bottom-dwellers. They will move out in autumn and back toward shore again in spring. Every few days they must find a new place to hide and to defend as a makeshift home. Each night, after venturing forth to feed, they either return home or travel on in the direction dictated by the season.

For any lobster, the twenty-fifth molt brings important changes. For the first time the animal is mature. Its length, measured along the barrel from the frontmost spine tip to the beginning of the tail, is likely to exceed three-and-nine-sixteenths inches. This is the legal minimum size at which a lobster may be kept by a person who finds it in his or her trap. It is a "counter"—a lobster the person can count on selling. The animal may be a male that has mated once, or it may be a female who has mated but not yet had time to have her first batch of eggs develop fully inside her body. Only if she has managed to get the eggs laid and transferred to the swimmerets beneath her tail, is she too precious for a lobsterman to keep. Such a female is in the "berry" condition, and must be tossed back to complete her maternal duties. Her young, developing inside the eggs she carries so obviously, keep her safe from August, when she expels her eggs, until the following June, when they hatch.

It seems strange that the home of the hatchling lobster is so unlike that of its parents. The young crustacean has a little more than a week in June to swim fearlessly near the sunny surface of the ocean. It uses its flat, black eyes in pursuit of minute creatures and rushes after them so vigorously that it catches a great many. But some of the little lobsters become food for herring that travel in vast schools. Each herring filters the water that enters its mouth and swallows anything the size of a young lobster without taking time to see what it is. The filter in the herring's mouth could not hold the particles of nourishment the little lobsters eat. The herring benefits from this food secondhand, by swallowing the lobsters that have caught it and are digesting it.

Many little lobsters survive to become lobsterlings, at which point they are heavier and properly equipped for their second way of life. They become stealthy night scavengers upon the ocean floor. Avoiding sunlight, they search by touch and taste. Nevertheless, only a few of the hatchlings that a mother lobster releases into the world succeed in scavenging for many years without getting caught. A male that lives to molt a thirtieth time will be fourteen years old and measure slightly more than nineteen inches in length. A female that achieves this size will be at least four months beyond her twentieth birthday. To reach the record length of thirty-four inches, a lobster would need to molt about twenty-eight more times. No one knows for sure if a male lobster that grows this big would be forty-two years old, or a female seventy-six. This is possible, even probable. The improbable feat is for the crustacean to escape the dangers of its second life for so very long a time.

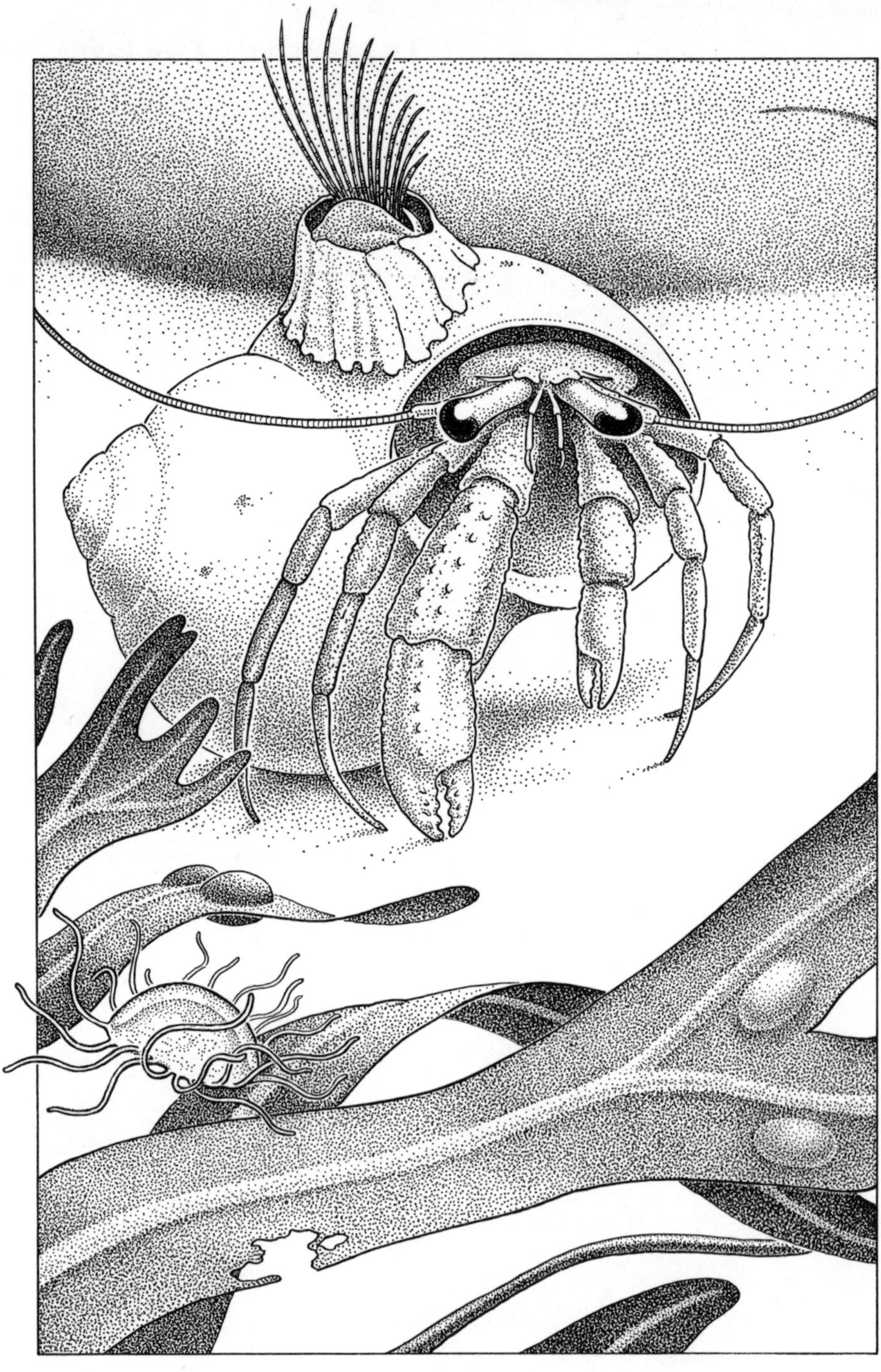

Hermit crab *(Pagurus)* and Acorn barnacle *(Balanus)*

10

The Hermit Crab and the Acorn Barnacle

THE MAGIC OF springtime affects the sea as well as the land; yet few people are sharp-eyed enough to notice when the coastal waters turn green. The change occurs just when sunny days help the new grass grow on front lawns, in city parks, and along country roads. Young sea animals of myriad kinds benefit from springtime changes. They float to the ocean's surface from greater depths, where they recently were hatched from eggs their parents had been holding. In surface waters the hatchlings have a wonderful chance to feed on microscopic plants—the ones that turn the water green.

Green plants drifting in surface waters are most numerous in spring. They multiply because the water around them has been fertilized by the great recycling processes of nature. Nitrates, phosphates, and carbon dioxide are released by bacteria all through the winter season. Their wastes are just what the green plants need to grow.

Every type of animal that inhabits salt water seems to send its young swimmers to the surface. Those that cannot live on green plants eat one another. Hatchling fishes, sea worms, mollusks, sea urchins, and sea stars can be found. Usually the hatchling crustaceans are the most numerous; sometimes they are called the "insects of the sea." They arrive in fantastic numbers and in many different forms. The surface waters teem with them. Seen through a strong lens, the water swirls with these minute living things like the tiny letters in freshly stirred alphabet soup. It is a trick to recognize even the most common types.

Many of the very young crustaceans look like miniature crayfish. Each has a few pairs of swimming legs under the large front part of its body and a jointed tail that it can curl under to propel itself backward. This bit of life is properly called a "zoea" (pronounced zoh-ee′-ah), from the Greek word for life. A zoea grows up to become a lobster, crab, or some less familiar creature; yet for most of the time the zoea feeds in surface waters, it is impossible to tell what it will grow up to be. Its appearance offers no clues. Among the little crustaceans that drift about in this way, one of the most remarkable is the hermit crab.

The zoea of a hermit crab may measure no more than a tenth of an inch in total length when it first begins swimming in search of food. It sheds its skin four separate times before its body gets twice as long. After the first molt, it can move its eyes separately from the rest of its body. The second molt lets it curl its tail tip. The next step in growth gives it paired swimmerets beneath its tail; these hang down and help propel the creature's body through the water, a task for which it previously used only its paired mouthparts and short legs.

After about three weeks, the hermit-crab zoea is ready to

molt again. This time it changes shape so drastically that for many years scientists had no idea what happened. The zoea becomes so unlike its former self that it used to be classified as an entirely different creature. Now we know that the zoea becomes something special called a "megalops," from the Greek word meaning gigantic eyes, before it can transform itself into a hermit crab. Its eyes seem much bigger than they were before because they are blacker and are held wide apart on movable stalks. The megalops has small pincers on its first pair of legs, which are still rather short. Its second and third pairs of legs have grown much longer and stick out to the sides. Its fourth and fifth pairs of legs remain short also. This is how the megalops first reveals that it is destined to become a hermit crab.

As a megalops, the little crustacean has a larger mouth and appetite than it had before. It rushes around in pursuit of drifting creatures that are still small enough to feed on microscopic plants. Fewer of these creatures are swimming about than a week or two earlier. Many of the small ones have grown bigger; others have been devoured. Only a lucky megalops gets so much to eat within two weeks that it is ready to molt, grow, and change shape once more. Even the green plants upon which all of the minute animals depend are getting scarcer. Quantities of them have been used as food, and the uneaten ones no longer can regenerate the supply because the fertilizing substances have been exhausted.

Clearly it is time for any full-grown megalops to travel somewhere else. The next molt provides the opportunity. Again the creature's body form changes abruptly. Now the young hermit crab has very long antennae. Its eyestalks, longer and more slender than before, turn forward, parallel to one another; they help the animal look ahead more than in any other direction. Its claws are bigger and stronger, particularly the one on the right side. The first three pairs of legs and the part of the body to which they are attached become strongly armored. Yet the rest of the body—the abdomen—stays very soft. The animal is longer than before, larger in diameter, almost as cylindrical as a worm, and somewhat twisted. It lacks swimmerets on the right side and has no jointed projection on the left side near the tail.

The little hermit crab sinks to the bottom and hunts almost frantically in all directions; yet it shows no interest at all in food. Miraculously, it hunts for something it has never seen or touched: an unoccupied snail shell. Generally it finds and discards a dozen before discovering one that is the right size. Some are much too big, others too small. The hermit crab feels each one with its antennae, then lifts and turns it with its pincers while standing or walking on its two other strong pairs of legs. It tests any empty shell that is close to the size needed to cover its soft abdomen by shoving that part of its body into the opening of the shell.

Almost every snail shell spirals clockwise as seen from its pointed end. Its cavity curves in exactly the right direction to accommodate the twisted abdomen of the hermit crab. A shell of the proper size even has space for the armored parts of the crab. The creature can pull itself in, using its short fourth and fifth pairs of legs, until only the tips of its claws are visible—and perhaps the ends of its antennae. The one jointed projection beside the end of the hermit crab's tail aids in shifting the body in or out and in holding tightly to the inside of the shell.

With an adequate shell as a portable home, the hermit crab is free to travel. It hikes along the bottom, searching for things to eat. It is primarily a scavenger. Any bit of meat becomes a prize, and pieces of seaweed will serve in an emergency. A soft-bodied worm will be prey if it cannot get away in time. Some hermit crabs crawl onto land and explore as far as a hundred yards from the nearest water. They hide in holes by day and emerge to feed at night. Only once or twice a week must they go back to the seashore to wet their bodies. This cleanses and freshens the soft skin of the abdomen inside the snail shell. Through this skin it exchanges carbon dioxide for oxygen.

No hermit crab ever seems satisfied with the shell it has. It runs over to any other hermit of similar size that it meets. The two touch feelers, and claws perhaps, and each examines the other's shell. Maybe one crab is a trifle larger than the other and wears a shell that is too tight. If the smaller crab has too loose a shell, the two may simply exchange. Both will be more comfortable for a while.

Every empty shell that a hermit finds is tested for size. In the course of a month the hermit may exchange its trailer home for a different one a dozen times. Sometimes the crab plays a trick on a beachcomber who is trying to make a fine collection of sea shells. A fresh trophy is left on an anthill so that the insects will clean out every bit of smelly meat inside. If the empty shell is left there overnight, the ants retreat into their nest underground. If a hermit crab comes along and finds the cleaned shell, it may carry it off as a new home, leaving behind a slightly smaller and worn shell. Rarely is the exchange one that the beachcomber appreciates.

Even a huge shell will appeal to a larger hermit if left in shallow water. Some of the hermit crabs along Florida coasts attain a length of ten inches or more. They can haul away discarded conch shells from beneath a dock, even those weighing as much as seventy ounces. Buoyed up by water, the shell is not quite so heavy.

The most spectacular of hermits on land is the coconut crab of the South Pacific. As a youngster it makes use of whatever snail shells it can find. Eventually it outgrows the largest that are available. The crab continues on its own, exploring islands fringed with coconut trees. The soft parts of the animal harden and it begins climbing the palms to feast on coconuts.

For other hermits, almost any snail shell of the right capacity is a fine fortress inside which the crab is safe. Yet from time to time the crustacean outgrows its own covering. It must molt to grow. During this molt, the hermit has to stay outside its snail shell, unprotected and vulnerable for almost a day, until its new body armor hardens. It tries to hide. Afterward, it must travel around naked, searching for a larger shell because the old one no longer fits. A fish that would not swallow a shell to get the crab inside, will gobble down the crab without its home. On land, a raccoon or other prowler may devour the freshly molted hermit as a tasty snack in a night of foraging. It is a lucky hermit that finds new shelter quickly, meets a mate, and later renews the numbers of tiny zoea in the sea.

Near any rocky coast, the hatchling hermit crabs swim side by side with tiny crustaceans of many other kinds. They pass one another again and again, with no recognition that later in life two of them may rendezvous. The hermit may carry a snail shell on which the other crustacean has attached itself. The passenger will be fixed for life, with no control over where it is going. Yet it benefits from being hauled to where there is water rich with food.

The passenger starts out as a minute creature. It and its close relatives are most numerous where the tide washes back and forth and the waves crash against the rocks. Unlike most crustaceans, it has a triangular body, all in one piece. It feeds itself and swims feebly by waving just three pairs of jointed, flexible appendages. These are its first and second pairs of antennae and its mandibles, or jaws. Its appendages cannot help the tiny crustacean to dodge any larger animal that is about to eat it. Incredible numbers are devoured by small fishes. The crustacean does discourage some pursuers with its sharp tail spine and others with substances that are either distasteful or poisonous. These substances are secreted by special glands through openings in the right and left front corners of the creature's body.

The paired, jointed organs with which the animal feeds on microscopic green plants identify this denizen of the sea surface as a crustacean. Yet this information about the hatchling does not go far enough, since more than 26,000 different crustaceans are known in their adult form. Identifying the creature as being in the "nauplius" stage, from the Greek word for a shellfish of some kind, does not answer the question: What will it be when it grows up? The answers are many. Some crustaceans that begin with a nauplius stage get no bigger than a grain of wheat; they spend their lives dancing on long antennae, searching for green plants near the surface of the sea. Others proceed through unimaginable changes to become permanent residents where the waves dash hardest against coastal rocks. A few take up residence on a shell that a hermit crab is carrying; these grow into barnacles.

A barnacle nauplius seems to be in no special hurry. It sucks in and digests the single-celled green plants that drift all around it. Every few days it gains enough weight to need a larger covering. It molts, expands quickly, hardens a new transparent shell, and eats some more. Within two weeks it will have shed its covering five times and grown long bristles on its paired appendages. It still will be seeing in almost all directions with a pair of compound eyes and an extra simple eye between them. Then comes the time to fast and to transform itself into something totally different.

Just as before, the nauplius sheds its thin, transparent armor, but it emerges clad in a bivalved shell. Only its first pair of antennae extend beyond the shell. Its compound eyes have drawn together into a single mass, its mouth is sealed shut, and its inner organs are being reorganized. It has six pairs of short, delicate legs. With them it pushes water through the crack between the two halves of its shell and slowly propels itself along.

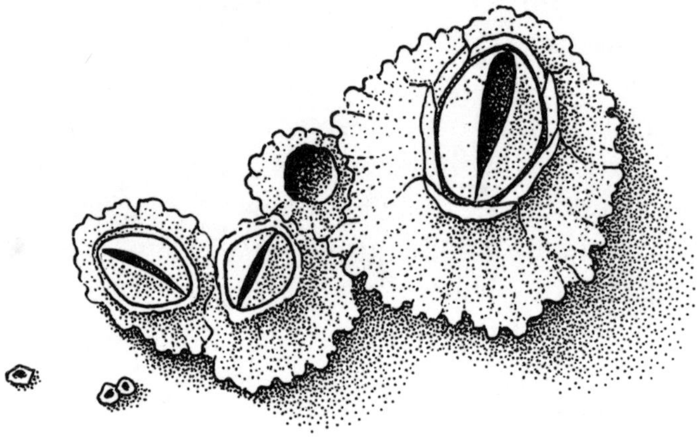

Acorn barnacles *(Balanus* spp.)

At this seventh stage in a barnacle's life, the creature has but one aim: to find a place to live for the rest of its existence. It shows a definite preference for rough surfaces and tiny cracks; so the worn shell that a hermit crab carries is fine. Wherever the site seems right, the animal cements its outstretched antennae to the support and no longer can move away. Its antennae shorten, bringing the small body firmly against the attachment spot. More cement is added. Known now as a "pinhead barnacle" because of its small size, the crustacean begins to secrete a limy fence around itself. Usually the fence consists of six sloping plates fitted together like the sides of an Indian's tepee. The fence ends squarely at the top, instead of rising to a pointed peak. An adjustable roof completes the limy shelter. It consists of four movable plates which can be slid apart to open the roof or drawn together to close it tight. The little barnacle seems safe inside. Its conical shape and firm attachment earn it the name "acorn barnacle."

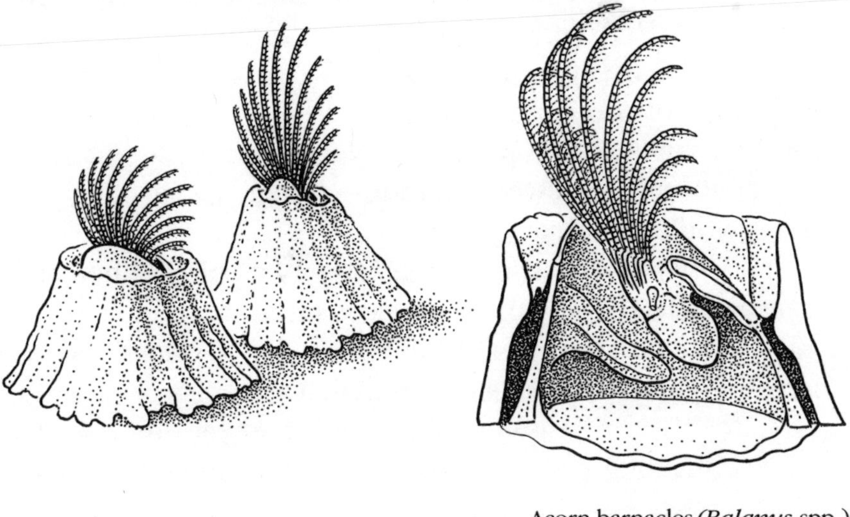

Acorn barnacles *(Balanus* spp.)

When sea water covers the barnacle's shelter, the animal quickly opens its roof doors. It reaches out with long, curved bristles on at least four pairs of feet, sweeping them through the water. It combs out particles of food and carries them to its reopened mouth. So fast is this action that the acorn barnacle can repeat it 140 times a minute if the surging water allows this opportunity. The distinguished biologist Louis Agassiz used to tell his students a century ago that an acorn barnacle is "nothing more than a little shrimplike animal, standing on its head in a limestone house and kicking food into its mouth." He might have added that this amazing creature detects any shadow that might signal danger, despite the fact that it has either absorbed its eyes or shed them.

Barnacles fascinated Charles Darwin as a young man. He studied barnacles from all over the world and wrote two large volumes about them. His books were published in 1851 and 1854 and established Darwin's reputation as a brilliant scientist. Despite all the studies made of them, these peculiar crustaceans, which settle for a permanent home so early in life, continue to conceal many a secret. No one yet has learned the ingredients of the extraordinary cement with which they attach themselves to a support. If dentists could use this material, repairing teeth would require much less drilling, and fillings would stay in place forever.

Acorn barnacles never reveal how they enlarge their limy shelters while keeping the wall plates fitted together so tightly. One type that Darwin discovered grows as much as four inches in diameter and six inches high. Daily measurements of its shelter show that it gets bigger at a fairly steady rate. The barnacle inside grows only when it molts. Somehow, while it holds its roof doors shut tightly, the creature sheds its covering like a body stocking. It gets one step bigger inside its home, waits for its new armor to harden, then opens the doors and kicks out the old suit. At this time it resumes feeding.

Rarely is an acorn barnacle alone. Any surface suitable for one usually attracts others. It may be a rock, a wharf piling, a ship's bottom, a sea turtle's shell, or even the skin of a giant whale. Competition for space and crowded conditions often limit the number of barnacles that can grow side by side. A life span of five years seems the most that can be expected in the wild, for so dense a population attracts hungry animals of various kinds. Dogwinkles and other snails drill holes through the limy armor and eat the acorn barnacle inside. Some sea stars and sea urchins find ways to gain entrance. A sheepshead or other fish with hard-rimmed jaws may crunch right through the strongest limy armor and soon clean off every barnacle from a rock or piling. This leaves a space where a fresh population of pinhead barnacles can settle the following spring.

Fate deals differently with the tiny, young crustaceans that swim so actively in surface waters of the sea. The secrets of an acorn barnacle include none that give it a long life, despite its ability to lock itself inside its own limestone stronghold. Hermit crabs have been known to survive three times as many years and to learn to accept food from a human hand. Alertness and a series of salvaged shells are the hermit's keys to continued success. The same shell, perhaps with a barnacle or two attached, can shelter the soft parts of one hermit crab after another. Each outgrows the shell and ignores the barnacles that ride along as nonpaying passengers. The barnacles know nothing of the crabs that haul them.

Even the most unlikely sites are soon occupied by animals, as members of each new generation seek the special opportunities they are equipped to enjoy. The suitable space may be only a bared area of rock where storm waves crash with enormous force, or it may be an empty snail shell resting on the sea floor, partly hidden by sand. It will not go unused for long. Nothing is wasted. Everything, live or dead, down to the last molecule, is recycled.

Afterword

ALMOST EVERYTHING IMPORTANT to an animal relates in some way to its home. Home has to be a familiar place, large or small, where the animal can rest. It must offer shelter of some kind from wind or wave, from disturbance by other animals, from the heat of the hot sun, the chill of cold nights, and the changes in weather that exceed what is seasonable. And home must be within reach of suitable foods.

In ancient Greece the word for home was *oikos*. In modern times this word has reappeared in English as the root of "ecology"—the study of the home life of animals and plants. Ecologists learn what is in the environment of each living thing. The environment supplies the energy and raw materials for growth and reproduction.

Some animals travel farther in their environment than others. And some move their shelters with them. Others have several homes from which to choose. The gadabouts succeed in special ways, and the stick-at-homes succeed in others. No one way is best. The world thrives on marvelous diversity.

The diversity that animals display in their behavior, whether they stay within a limited territory or wander widely, provides one immense benefit. It reduces to the minimum their competition for food and space. It permits more kinds of creatures to hold a place on earth than otherwise would be possible.

Much of this diversity lies just beyond our doorstep. As Gilbert White, an observant naturalist of Selbourne, England, wrote on October 8, 1768, ". . . all nature is so full, that that district produces the greatest variety, which is the most examined."

Index

Index

B

C

Index

T

Index